T5-CPZ-883

THE ESTUARINE ENVIRONMENT

The inaugural symposium of the
Estuarine and Brackish-water Biological Association
(*now renamed the Estuarine and Brackish-water Sciences Association*)
held on 13*th October*, 1971

THE ESTUARINE ENVIRONMENT

Edited
for the Association
by

R. S. K. BARNES, Ph.D.
Hon. Secretary, Estuarine and Brackish-water Sciences Association
and
Professor J. GREEN, D.Sc.
Department of Zoology, Westfield College, University of London

APPLIED SCIENCE PUBLISHERS LTD
LONDON

APPLIED SCIENCE PUBLISHERS LTD
RIPPLE ROAD, BARKING, ESSEX, ENGLAND

ISBN 0 85334 539 2

WITH 11 TABLES AND 33 ILLUSTRATIONS

Printed in Great Britain by Galliard (Printers) Ltd Great Yarmouth

Contributors

Professor DON R. ARTHUR
Department of Zoology, King's College, University of London

M. J. BARRETT
Water Pollution Research Laboratory, Stevenage, Hertfordshire

Professor L. C. BEADLE
Department of Zoology, University of Newcastle upon Tyne

Dr G. M. DUNNET
University of Aberdeen Culterty Field Station, Newburgh, Aberdeenshire

Dr K. R. DYER
Department of Oceanography, University of Southampton (*Present address: NERC Unit of Coastal Sedimentation, Taunton, Somerset*)

Dr R. L. JEFFERIES
School of Biological Sciences, University of East Anglia, Norwich

Dr H. MILNE
University of Aberdeen Culterty Field Station, Newburgh, Aberdeenshire

JOHN PHILLIPS
CERL Marine Biological Laboratory, Fawley, Hampshire

Professor W. D. P. STEWART
Department of Biological Sciences, University of Dundee

Dr P. R. WALNE
MAFF Fisheries Experiment Station, Conway, Caernarvonshire

Contents

Prologue

THE ESTUARINE AND BRACKISH-WATER SCIENCES ASSOCIATION

The Association, founded at the symposium recorded in this volume as the Estuarine and Brackish-water Biological Association, is a scientific body with a sphere of interest encompassing all aspects of the biology, chemistry, geology, hydrography and human use of aquatic environments other than those of fresh water and of the sea. Environments that therefore fall within this sphere of interest are those subjected to fluctuating salinities on the one hand, for example estuaries, drainage-ditch systems and salt-marshes, and those of more constant salinity on the other, for example brackish seas and lagoons, which are usually hyposaline to normal sea water, and inland saline waters, which are often hypersaline. Although it was founded in the United Kingdom, it is hoped that the Association will progressively become world-wide.

The objects of the Association are to 'encourage the production and dissemination of scientific knowledge and understanding concerning estuaries and other brackish waters' and to promote co-operation, co-ordination and communication between the producers and the users of scientific information and between specialists of different disciplines having a common interest in brackish waters. This it expects to achieve by holding interdisciplinary meetings on subjects of importance, by causing to be published handbooks and papers on subjects of particular interest, and by keeping members informed of developments in estuarine science throughout the world. Other activities will be undertaken if the need arises.

The Association is controlled by a Council consisting of representatives elected by the membership and persons nominated by sponsoring organisations. Membership is open to anyone interested in furthering the aims of the Association.

Preface

This small volume records the series of invited papers given at a symposium on 'The Estuarine Environment' held in the Meeting Rooms of the Zoological Society of London, Regent's Park, on the 13th October 1971. This gathering was occasioned by the inauguration of a new scientific body, provisionally entitled the 'Estuarine and Brackish-water Biological Association'.

It is to be hoped that this book will do something towards correcting the general lack of research information on the subject of British estuaries gathered in a single volume. Estuaries are complex environments existing at the borders of the sea, rivers and the land. This admixture of environmental influences results in complex interdisciplinary problems and these require interdependence of scientists from many different backgrounds for their successful solution. Therefore, reflecting this necessity, the papers in this volume cover a wide field, including areas lying within at least four disciplines, namely zoology, botany, chemistry and geology. Each paper is concerned with an important growth point in estuarine science. Regrettably, but unavoidably owing to the limited time available, a number of equally important fields of study had to be excluded from the symposium. Perhaps the following papers will at least stimulate the interested reader to explore the areas neglected by this volume.

As part of the task which this Association has set itself concerns the promotion of effective communications between the various disciplines involved in estuarine research and also between the producers and users of the scientific information obtained, it was felt at this symposium that the inclusion of the name of a single discipline in the Association's title was not appropriate. Accordingly, the Association has been renamed to more accurately indicate its field of interest.

The Association would like to record its gratitude to the Fishmongers' Company, Unilever Ltd, Imperial Chemical Industries Ltd, Esso Petroleum Ltd and the Zoological Society of London for enabling by their generosity this symposium to be held; to Professor R. W. Edwards and R. S. Glover Esq. for chairing the meeting; and to G. B. Olley Esq. and the publishers for their interest in making the papers available to a wider audience. It only remains to thank the individual contributors for making the task of the editors a light (and pleasant) one and to thank those attending the symposium for making it such a success.

THE EDITORS

1

Estuarine and Brackish Waters—an Introduction

W. D. P. Stewart

'Its consideration, therefore, is certain, if properly entered into, to be fruitful of interesting and valuable thought.'

J. W. DAWSON, 1875

My role here is to try to provide a brief introduction to the symposium proper which follows. This is a tall order when one appreciates the enormity of the subject under discussion. For a start, the term 'estuary' is difficult, if not impossible, to define precisely. In Lauff (1967) seven different definitions are given in the first seven pages. Nevertheless, it is fairly well accepted in general terms at least that estuaries are areas in which sea water is appreciably diluted by fresh water from rivers, while brackish waters are any waters in which sea water is diluted by fresh water irrespective of the origin of the latter. Thus estuaries are brackish water habitats whereas brackish water habitats are not necessarily estuaries. More than half of the British coastline is subjected to estuarine or brackish waters and none of us live more than about fifty to sixty miles from such waters. It is not surprising therefore that the United Kingdom, with one of the largest brackish habitat: land mass ratios in the world, should be the birthplace of a new organisation having estuarine and brackish habitats as its prime interest. What is surprising perhaps is that it is only now that such an association is being initiated.

Because of the high population density of the United Kingdom and in particular the fact that we are an island, increasing pressure is continuously being brought on our estuarine and brackish areas. Many of our major cities have developed on rivers and estuaries: London, Glasgow, Liverpool, Southampton, Bristol, Edinburgh, Dundee, Hull, etc. These provide important links with the Continent and with the world at large and increasing population density and industrialisation in these areas are providing problems not encountered hitherto in our islands. In addition, as populations increase, greater attention is being given to reclaiming from our estuaries good land suitable for industrial development, urbanisation and agriculture. At present we have the Morecambe Bay barrage feasibility scheme, the findings of which are awaited with interest, and the possibilities of reclaiming additional land on a large scale from the Wash and from the Thames are now being considered seriously. It is essential, therefore, that an attempt should be made to co-ordinate the obvious interest that there is in many quarters in estuarine research. It is equally important that this interest is sustained. Estuarine and brackish habitats should not be allowed to develop into a no-man's land to be considered in detail neither by fresh-water biologists nor by oceanographers. Fortunately in Britain our marine and fresh-water laboratories have appreciated the importance of estuaries and active research is being carried out by these and by other groups. In addition the new Institute for Marine Environmental Research at Plymouth will be engaged in this field. Figure 1 shows some of the areas where estuarine research is being carried out and it is obvious that there is an active interest in most regions of the country.

Current interests in estuaries range from the compilation of purely descriptive lists of plants and animals (an important prerequisite if future trends of our estuaries are to be followed) through to studies on the role of micro-organisms in accretion, erosion and land reclamation. There is interest in the effects of industrialisation, such as the possible expansion of the petrochemical industry on the Scottish east coast, the problems of using rivers as waste disposal systems and the effects of agricultural drainage on our rivers and estuaries. There is active interest in the importance and control of *Spartina anglica*, and the question of keeping our navigation channels open. Some of these problems will be considered by succeeding speakers, and I should like to take the opportunity here to try to emphasise briefly three of the ways in which interactions in estuaries and brackish waters may affect marine, terrestrial and fresh-water ecosystems.

As an example of a way in which estuarine areas can affect the truly marine environment we can consider the problem of vitamin B_{12} production. Estuarine areas are particularly rich in vitamin B_{12}, a growth factor which is essential for the growth of many of the marine micro-algae which are important primary producers in our seas and oceans. Evidence of a vitamin B_{12} requirement for many such algae was established, particularly during the 1950s and subsequently, by the work of Provasoli and his collaborators at the Haskins' Laboratory, New York, by Droop at the Scottish Marine Biological Association and by others (*vide*, *e.g*. Droop, 1962;

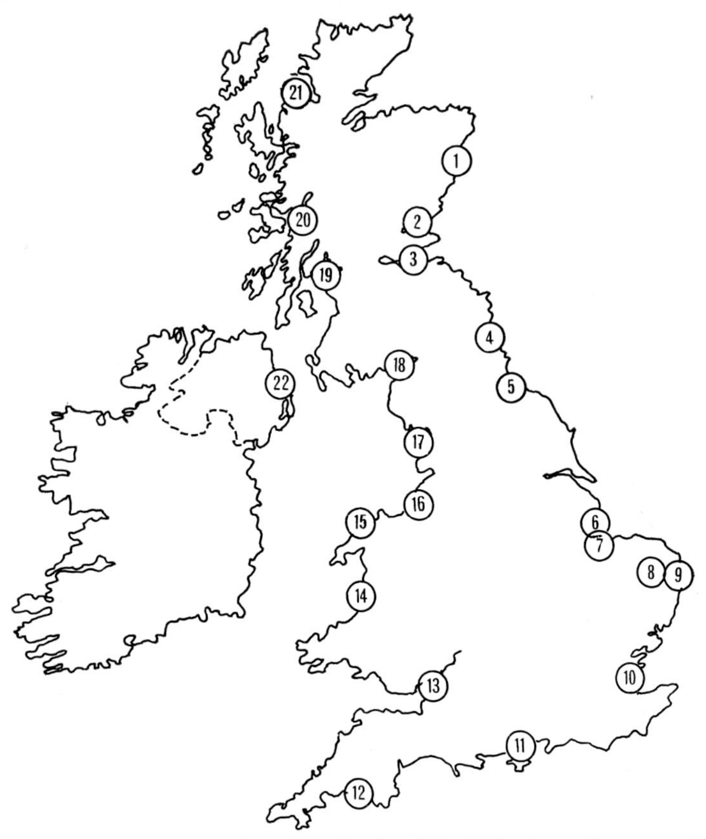

FIG. 1. Some areas and laboratories in the United Kingdom where estuarine and brackish water research is being carried out.

1. Dept Zoology, University, Aberdeen (Ythan Estuary); Dept Agric. Fish. Scotland; 2. Dept Biol. Sci., University, Dundee; Tay Estuary Research Centre, University, Dundee (Tay Estuary); 3. Dept Brewing Biochem., Herriot-Watt University, Edinburgh; University of Edinburgh; Freshwater Fish. Lab., Pitlochry (Forth Estuary); 4. Dept Zoology, University, Newcastle-upon-Tyne (Tyne); 5. Dept Zoology, University, Newcastle-upon-Tyne (Tees); 6. Dept Botany and Zoology, University, Nottingham (Gibraltar Point); 7. Inst. Mar. Environ. Res., Plymouth (Wash); 8. University East Anglia; Nature Conservancy (Norfolk estuaries); 9. Min. Agric. Fish. Food, Burnham-on-Crouch and Min. Agric. Fish. Food, Lowestoft; 10. Dept Zoology, King's College, and Dept Zoology, Queen Mary College, London (Thames); 11. Dept Oceanography, University, Southampton; Central Electricity Research Lab. Fawley (Southampton Water); 12. Marine Biological Association, Plymouth (Tamar), Inst. Mar. Environ. Res., Plymouth; 13. Project Sabrina, University, Bristol (Severn); 14. Dept Zoology, University College, Aberystwyth (Dovey); 15. Marine Station, Menai Bridge (Menai Straits), Min. Agric. Fish. Food, Conway; 16. Liverpool Bay Group, University, Liverpool (Liverpool Bay); 17. Inst. Mar. Environ. Res., Plymouth (Morecambe Bay); 18. Dept Biology, University, Strathclyde (Solway Firth); 19. Clyde Riv. Purif. Bd; University, Strathclyde; Freshwater Fish. Lab., Pitlochry (Clyde); 20. Scottish Marine Biol. Assoc., Oban (Loch Etive); 21. Dept Agric. Fish. Scot., Aberdeen (Loch Ewe); 22. Dept Zoology, University, Belfast (Belfast Lough).

Provasoli, 1963). In addition to vitamin B_{12}, thiamine and biotin are sometimes required as well, but less frequently (Table 1).

Simultaneous with the advances being made in characterising the vitamin requirements of pure cultures of algae other workers such as Starr (1956) and Burkholder and Burkholder (1956) were measuring vitamin B_{12} contents of estuarine muds and of detritus washed out

TABLE 1

VITAMINS REPORTED AS BEING REQUIRED BY ALGAL STRAINS OF DIFFERENT ALGAL GROUPS

(from Droop, 1962)

				Strains which require				
Division	*Total*	*Auxotrophs*	*Non-auxotrophs*	*Only vitamin B_{12}*	*Only thiamine*	*Vitamin B_{12} and thiamine*	*Thiamine and biotin*	*Vitamin B_{12}, thiamine, and biotin*
Chlorophyta	47	22	24	10	7	5	0	0
Euglenophyta	10	10	0	0	1	9	0	0
Cryptophyta	11	11	0	2	1	8	0	0
Pyrrophyta	17	17	0	12	0	0	1	4
Chrysophyta	13	12	1	2	1	6	1	2
Bacillariophyta	54	21	33	11	6	4	0	0
Phaeophyta	1	0	1	0	0	0	0	0
Rhodophyta	1	1	0	1	0	0	0	0
Cyanophyta	25	1	24	1	0	0	0	0
Totals	179	95	83	39	16	32	2	6

from estuaries. Starr (1956) showed in studies on salt-marsh areas near Sapelo Island in Georgia that the water draining out from salt-marshes was particularly rich in vitamin B_{12} and the concentrations, which were correlated directly with levels of detrital material, were lower as one moved away from the marshes towards the more open oceanic waters (Fig. 2). In estuarine and salt-marsh areas vitamin B_{12} is produced primarily by the prokaryotic bacteria and blue-green algae and in the open seas heterotrophic bacteria are probably the more important sources of the vitamin. Nevertheless it is apparent from the studies of Starr and of others that estuarine areas can be important sources of vitamin B_{12}, and other nutrients as well, for coastal algae in particular.

A good example of the effect which this may have is the well-documented blooms of the 'red tide' dinoflagellate *Gymnodinium breve* which occur off the west coast of Florida and in the Gulf of Mexico where waters of salinity 32–34$^0/_{00}$ predominate (Ketchum and Keen, 1948). *Gymnodinium* requires in the bacterised state vitamin B_{12}, soil extract and a chelating agent for healthy growth

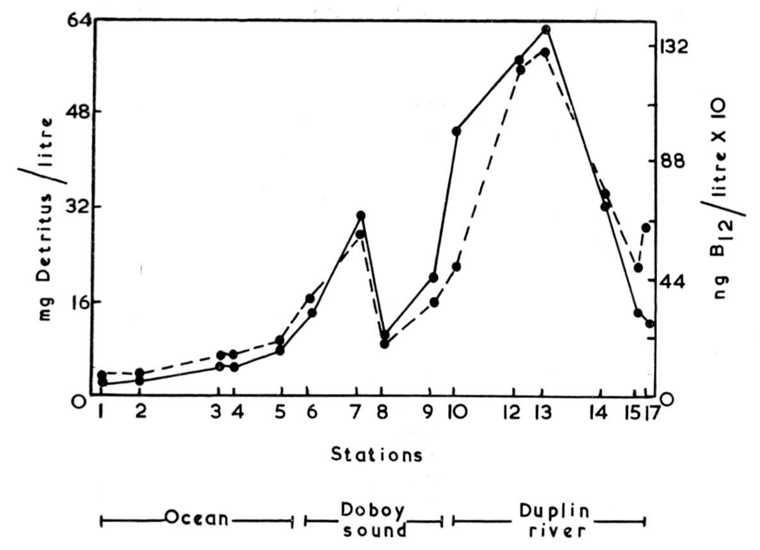

FIG. 2. Correlation between vitamin B_{12} levels (●—●) and level of detritus (●- - -●) near Sapelo Island, Georgia (after Starr, 1956).

(Wilson and Collier, 1955) and as Provasoli (1958) remarks these are exactly the factors which water from estuaries and run-off from land would provide in marine coastal waters. A rather similar situation exists in the southern Atlantic Ocean where highly productive waters occur hundreds of miles off shore from the mouth of the Amazon.

Secondly, there is the importance of micro-organisms in the process of accretion which affects the development of sand bars, salt-marsh development and sand-dune build up and which thus is of importance to conservationists, terrestrial ecologists and those interested in maintaining open navigation channels. Bacteria, fungi, and algae all colonise sand grains in estuarine and brackish habitats and thus help in the initial aggregation process. The importance of algae in salt-marsh and in estuarine areas was in fact appreciated by early workers such as Cotton (1912) on Clare Island and by Carter (1932; 1933a, b) on the Ynyslas and Canvey marshes. In these and other marshes the intertwining filaments of organisms such as *Microcoleus chtonoplastes, Hydrocoleum lyngbyaceum* and others which penetrate below the marsh surface, particularly in autumn, winter and spring, have an important binding effect. Some of the algae in salt-marsh and dune slack areas also fix gaseous nitrogen from the atmosphere (Stewart, 1965; 1967a) and studies using ^{15}N as tracer show that this fixed nitrogen becomes available for the

TABLE 2

ACCRETION OF NITROGEN AND NITROGEN FIXATION BY *HIPPOPHAË RHAMNOIDES* L. AT GIBRALTAR POINT, LINCOLNSHIRE
(Stewart and Pearson, 1967)

Mean age of plants (years)	*Mean total N accreted*[a] *(kg/ha)*	*Mean total N fixed*[a] *(kg/ha)*
3	27·2	4·4
11	69·3	41·9
13	133·3	57·5
16	178·9	1·5

[a] The nitrogen accretion data refer to the mean total nitrogen accreted per annum by plants of 0–3 years, 3–11 years, 11–13 years and 13–16 years, whereas the nitrogen fixation data refer to nitrogen fixed by plants 3, 11, 13 and 16 years old.

growth of associated non-nitrogen-fixing organisms (Stewart, 1967b). In sand-dunes nitrogen-fixing micro-organisms on the surfaces of roots may provide nitrogen for plants such as *Ammophila arenaria* (Hassouna and Wareing, 1964) and actinomycete-like organisms in the root nodules of *Hippophaë rhamnoides* also help in the build up of soil by fixing nitrogen (Table 2). It is paradoxical that at present control measures against the spread of *Hippophaë rhamnoides* have been suggested. Fungi which are components of developing salt-marshes (*see* Pugh, 1962) also help in sand-grain binding, and ectotrophic mycorrhizae occurring in symbiotic association with higher plants are the predominant fungal component of coastal sand-dunes in Britain. There, they help in the accretion process by facilitating uptake of nutrients in general and phosphorus in particular (*see* Nicolson, 1967).

Thirdly, there is the use of rivers as waste disposal systems for urbanisation, industry and agriculture. With man more or less restricted to the terrestrial quarter of the Earth's surface this is probably the most satisfactory means of waste disposal, but there at least are two disadvantages to such a system which affect estuarine areas. One is that the waste products, including toxins and nutrients, do not necessarily move *in toto* into the open waters. We have found, for example, in the River Tay in Scotland that mercury disposed of higher up the river tends to accumulate in certain algae including species of *Ulva* and *Porphyra* which occur near the mouth of the estuary. Second, there is no guarantee that the waste products which are washed out into the open sea and which become deposited on the ocean bed actually remain there. If marshes and estuaries are accreting rapidly, due, for example, to few off-shore winds and strong on-shore currents, much of the material will return to shore by tidal mechanisms. Thus studies on tidal currents and movement should not be underestimated, not only in relation to simple waste disposal by rivers but also in connection with the dumping of materials at sea by means of dredgers, etc. The reports of Perkins and Williams (1966) and Perkins, Williams and Gorman (1966) on the disposal of the radioactive ^{106}Ru from the Windscale reactor into the Solway Firth emphasises some of the problems associated with waste disposal. These workers found that ^{106}Ru originally discharged into the Solway Firth and carried out to sea was subsequently carried back into parts of the Solway estuary adsorbed on to particulate material which accumulated as silt on salt-marshes and on rocky

TABLE 3

LEVELS OF ^{106}RU IN MUSSELS AND SOIL COLLECTED FROM BALCARY BAY, SOLWAY FIRTH

(Perkins *et al.*, 1966)

Month	*Specific activity ^{106}Ru*		*Apparent concentration factor*
	Soil (pCi/g dry)	*Mussel (pCi/g ash)*	
1962 June	44·0	334·0	7·6
July	22·5	390·0	15·3
August	34·0	294·0	8·6
September	45·0	443·0	9·8
October	29·0	438·0	15·1
November	120·0	404·0	3·4
December	113·0	350·0	3·1
1963 January	145·0	337·0	2·3
February	135·0	253·0	1·0
March	135·0	621·0	4·6
April	152·0	880·0	5·8

shores. As a result some of the local fauna such as *Corophium volutator, Macoma balthica* (Table 3) and certain plants such as the angiosperm *Spartina anglica*, and the algae *Pelvetia canaliculata* and *Fucus spiralis* showed higher than background levels of radioactivity. Perkins and Williams also found that, unlike the filter feeding animals, there was no significant accumulation of ^{106}Ru in the economically important plaice and flounder in the Solway Firth.

The examples which I have touched upon have been chosen because they provide examples of three financially important aspects of estuarine and brackish habitat research: factors affecting the distribution of primary producers in the sea, factors affecting accretion and erosion, and factors influencing the disposal of waste effluent from terrestrial and fresh-water ecosystems. Yet a fair proportion of the research covered in the first two aspects was conceived initially as pure scientific research with no particular economic or applied bias. This emphasises, I hope, the importance of basic research as well as applied research. It was Pasteur who said that there were no applied sciences, only science and the applications of science. It is hoped that the Estuarine and Brackish-water Sciences Association will help to provide a forum for the efficient collation and interchange of ideas and information between pure and applied scientists and also among industrialists, workers from water resources boards and indeed the general public. A great deal of information is available on estuaries at the moment but it is scattered throughout the literature, the papers have been read at diverse scientific meetings and as a result the current state of affairs in British estuarine and brackish waters is difficult to assess. The organisation of this symposium is the first step which the new Association has taken to rectify this. Like those people who are attending I look forward with interest to the future of the Estuarine and Brackish-water Sciences Association.

REFERENCES

Burkholder, P. R. and Burkholder, L. M. (1956). 'Vitamin B_{12} in suspended solids and marsh muds collected along the coast of Georgia.' *Limnol. Oceanogr.*, **1,** 202–8.

Carter, N. (1932). 'A comparative study of the alga flora of two salt marshes, part I.' *J. Ecol.*, **20,** 341–70.

Carter, N. (1933a). 'A comparative study of the alga flora of two salt marshes, part II.' *J. Ecol.*, **21,** 128–208.

Carter, N. (1933b). 'A comparative study of the alga flora of two salt marshes, part III.' *J. Ecol.*, **21,** 385–403.

Cotton, A. D. (1912). 'Marine algae. Clare Island survey.' *Proc. Roy. Irish Acad.*, **31,** 1–178.

Droop, M. R. (1962). 'Organic micronutrients.' In *Physiology and Biochemistry of Algae.* Ed. R. A. Lewin, pp. 141–59. New York: Academic Press.

Hassouna, M. G. and Wareing, P. F. (1964). 'Possible role of rhizosphere bacteria in the nitrogen nutrition of *Ammophila arenaria.*' *Nature, Lond.*, **202,** 467–9.

Ketchum, B. H. and Keen, J. (1948). 'Unusual phosphorus concentrations in the Florida "red tide" sea water.' *J. Mar. Res.*, **7,** 17–21.

Lauff, G. H. (Ed.) (1967). *Estuaries.* Washington, D.C.: Publication No. 83. American Association for the Advancement of Science, p. 757.

Nicolson, T. H. (1967). 'Vesicular-arbuscular mycorrhiza—a universal plant symbiosis.' *Sci. Prog. Oxf.*, **55,** 561–81.

Perkins, E. J., Williams, B. R. H. and Gorman, J. (1966). 'The biology of the Solway Firth in relation to the movement and accumulation of radioactive materials. V: Radioactivity in the fauna.' *United Kingdom Atomic Energy Authority* PG Report 752 (CC).

Perkins, E. J. and Williams, B. R. H. (1966). 'The biology of the Solway Firth in relation to the movement and accumulation of radioactive materials XI: General discussion.' *United Kingdom Atomic Energy Authority* PG Report 753 (CC).

Provasoli, L. (1958). 'Growth factors in unicellular marine algae.' In *Perspectives in Marine Biology*, Ed. A. A. Buzzati-Traverso. Pp. 385–403. Berkeley: University of California Press.

Provasoli, L. (1963). 'Organic regulation of phytoplankton fertility.' In *The Sea*, Ed. M. N. Hill, Vol. 2, pp. 165–219. New York: Wiley.

Pugh, G. J. F. (1962). 'Studies on fungi in coastal soils. II: Fungal ecology in a developing salt marsh.' *Trans. Brit. mycol. Soc.*, **45,** 560–6.

Starr, T. J. (1956). 'Relative amounts of vitamin B_{12} in detritus from oceanic and estuarine environments near Sapelo Island, Georgia.' *Ecology*, **37,** 658–64.

Stewart, W. D. P. (1965). 'Nitrogen turnover in marine and brackish habitats I: Nitrogen fixation.' *Ann. Bot., N.S.*, **29,** 229–39.

Stewart, W. D. P. (1967a). 'Nitrogen turnover in marine and brackish habitats II: Use of ^{15}N in measuring nitrogen fixation in the field.' *Ann. Bot., N.S.*, **31,** 385–407.

Stewart, W. D. P. (1967b). 'Transfer of biologically fixed nitrogen in a sand dune slack region.' *Nature, Lond.*, **214,** 603–4.

Stewart, W. D. P. and Pearson, M. C. (1967). 'Nodulation and nitrogen fixation by *Hippophaë rhamnoides* L. in the field.' *Plant and Soil*, **26,** 348–60.

Wilson, W. B. and Collier, A. (1955). 'Preliminary notes on the culturing of *Gymnodinium brevis* Davis.' *Science*, **121,** 394–5.

2

Sedimentation in Estuaries

K. R. DYER

INTRODUCTION

'An estuary is a semi-enclosed coastal body of water having a free connection with the open sea and containing a measurable quantity of sea salt' (Pritchard, 1952). Study of estuarine sedimentation, however, requires rather broader considerations as the major sources of sediment are generally outside the estuary. Consequently, knowledge of littoral and river processes are required. Inside the estuary sediment movement is related to the circulation patterns of the water. These are a complex interplay of ever-changing river and tidal flows complicated by the effects of topography.

Most estuaries are the result of the drowning of river valleys incised during the late Pleistocene low sea level period. Sedimentation in some has kept pace with the gradual inundation, but others may still be slowly adjusting to a new equilibrium, which is further complicated by man who commonly interferes in the production of natural balances. Estuaries are mainly floored by recent sediments and deposition and reworking of them is more active than the erosion of older rocks.

One difficulty remains paramount in dealing with estuarine sediments; it is that extreme conditions are often more important than the average. One flood of the river can discharge more sediment in days than normally appears in years. For instance, the Delaware River carried more sediment in two days in 1955 than during any other year between 1950 and 1966. Under extreme flow conditions the estuary, as defined above, may not exist and the deposits of long periods can be flushed into the open sea. Measurement of the parameters necessary for analysis of the sediment movement at such times, may be difficult if not hazardous.

That no two estuaries are alike discourages generalisations as one never knows whether general principles or unique details are being studied. However, many comprehensive studies allow some general conclusions to be stated. There is an extensive geological literature on recent sediment distributions in estuaries and on estuarine sediments within the geological column. There is also a similar engineering literature concerning studies on model rivers and estuaries. Here we will attempt to briefly review evidence on the processes involved in present day sedimentation in estuaries. Useful additional information is available in Lauff (1967), Guilcher (1963) and Shepard (1948).

ESTUARINE CIRCULATION SYSTEMS

Model studies have shown that the correct sediment distribution is produced only if salinity is reproduced as well as the correct river discharge, tidal flow and topography. The pattern of shoaling varies with the type of mixing and estuaries have been classified in this respect by Pritchard (1955). The roles of differing river flows and tidal prism volumes and of topography become apparent in such a classification.

Salt Wedge Estuary

When the river discharge is high compared with the tidal flow, the fresh water flows outwards on the surface, overriding a wedge of saline water. The upper surface of the wedge has a sharp halocline that can reach $20^{0}/_{00}$ in less than 0·5 m. Salt water is mixed upwards by the breaking of internal waves on the interface. To compensate for this loss of salt water there is a slow landward movement within the salt wedge. However, the salinity varies little from the wedge tip to the estuary mouth. River flow variations alter the distance of penetration of the salt wedge into the estuary. An example of a salt wedge estuary is the Vellar, described by Dyer and Ramamoorthy (1969). Sedimentation in these estuaries is usually dominated by river floods. Where sediment discharge is high deltas can form.

Partially mixed estuary

As tidal motion of the water mass becomes important, extra turbulence is generated by the water movement across the bottom. This exchanges salt water into the upper layer and fresh water into the lower layer, and salinity in both layers increases towards the mouth. The turbulent mixing increases the discharge of both upper and lower layers and the landward net flow near the bottom becomes appreciable. The intensity of this net flow is affected by river discharge variations (Simmons, 1955). This landward bottom flow causes sediment movement up the estuary and accumulation near the landward end of the saline layer (Meade, 1969).

In partially mixed estuaries the effects of Coriolis force become apparent leading to the development of a horizontal circulation pattern. Many estuaries of this type have a cross-sectional area that increases exponentially towards the sea, this may be the result of an attempted balance between deposition and erosion.

An example of the partially mixed estuary is the James River (Pritchard, 1954).

Vertically homogeneous estuaries

With increasing tidal range, mixing can become sufficiently intense for the water column to be homogeneous. Estuaries of this type are generally shallow. If there is lateral inhomogeneity then, under the influence of Coriolis force, the mean fresh-water flow is seaward at all depths on the right and landward at all depths on the left (in the northern hemisphere). The circulation system is now horizontal. With the laterally homogeneous estuary the horizontal effects are suppressed, there is seaward flow throughout the cross-section and upstream transport is by diffusive processes probably depending largely on the effects of the estuarine topography.

Fjords

Fjords are really a special case of a salt wedge estuary, but with a very deep, sometimes stagnant, lower layer.

These different types of estuary are really stages in a continuous sequence. As can be seen from their characteristics, different statification and circulation systems are produced by different ratios of river flow to tidal prism volume (Simmons, 1955). Hansen and Rattray (1966) have developed a quantitative classification system in which estuarine characteristics are measured in terms of a stratification parameter and a circulation parameter.

FLOCCULATION AND CONSOLIDATION

Most of the sediment in motion in estuaries is carried in suspension and is of small size. Those particles less than about 2 microns in diameter are mainly composed of clay minerals, principally illite, kaolinite and montmorillonite. They normally have a negative surface charge arising from broken intermolecular bands at the edges of the clay mineral lattice and from cationic substitution within the lattice. Each particle is surrounded by adsorbed anions. The total negative charge is balanced by a double layer of hydrated cations. The stability of this layer depends largely on the ion concentration in the surrounding water, and, as the ion concentration rises, there is a tendency for the clay particles to flocculate. Flocculation occurs when particles collide and the probability of this happening will be a function of particle concentration. Collisions can occur by Brownian motion and will be promoted by turbulent mixing if the suspension is sheared. Differential sedimentation of floccules will mean that the larger aggregates are more likely to encounter other particles and grow than the smaller ones. Under certain conditions little flocculation appears to occur at concentrations less than about 300 ppm.

With adequate particle concentrations, flocculation of illite and kaolinite is complete above a salinity of about $4^0/_{00}$. The size, and consequently the fall velocity, of the montmorillonite floccules, however, varies over the entire salinity range up to $35^0/_{00}$. Fall velocities for illite, kaolinite and montmorillonite in full sea water are 0·083, 0·0135 and 0·0015 cm/s respectively (Whitehouse *et al.*, 1960). For Portishead mud, flocculation was complete at a sodium chloride/mud ratio (g/g) greater than 0·0165 (Peirce and Williams, 1966). Though the individual particles may have a density of about 2·6 g/cm^3, the floccules may be much less dense, quoted values ranging between 1·27–1·8 g/cm^3. Consequently their size and effective surface area may be much larger than that of a quartz grain having the same fall velocity.

Whitehouse *et al.* (1960) also found that a temperature decrease from 26 to 6°C decreased settling rates by up to 40%. Flocculation was also affected by the presence of complex carbohydrates in concentrations between 0·0005–1 g/l, which increased size and fall velocities by up to 25% and some proteins which decreased fall velocities by up to 40%. In some estuaries the floccules may be organically formed as faecal pellets.

Flocculation is a reversible phenomenon. Floccules carried by the estuarine circulation into fresh water will disaggregate, and intense turbulence can also physically disrupt them. The presence of sewage and industrial wastes will obviously affect the processes of both flocculation and deflocculation.

If suspended sediment is present in sufficiently high concentrations turbulence can be damped and the floccules will not fall as separate units, but in layers. In many estuaries, especially at slack water, layering above the bottom is observed on echo sounder records. In the Thames these ‘fluid mud’ layers have particle concentrations of

100 000 ppm (Inglis and Allen, 1957) and in the Chao Phya estuary layers with concentrations up to 300 000 ppm can be 2·5 m thick (Allersma *et al.*, 1966). Under the influence of the tidal currents the mud moves as a turbid layer with concentrations of 2 000–20 000 ppm, close to the bottom. Einstein and Krone (1962) have examined the properties of suspensions of San Francisco Bay mud. They found that the suspension settled in two phases. The first phase, the 'fluid mud' phase, appeared to start at a concentration of about 10 g/l and settling continued by expulsion of interfloccule water until a concentration of 167 g/l and a density of 1·11 g/cm^3 was attained. In still water the fluid mud phase lasted about two hours. During the second stage the sediment consolidated slowly, presumably by expulsion of further interfloccule water and collapsing of the flocculate structure.

As consolidation proceeds, providing the sedimentation rate is low enough to allow the consolidation to keep pace with the increasing overburden pressure, there is a linear increase in shear strength and effective overburden pressure with depth. The moisture content decreases exponentially with depth. Under these conditions the sediment is normally consolidated and certain relationships between overburden pressure, shear strength, void ratio and Atterberg limits are predictable (Skempton, 1970). The Atterberg limits are empirical, but effective, indices that compare the physical characteristics of clays, such as plasticity, that are dependent on the mineralogy. The plasticity index is related to the percentage of the mud finer than 2 microns and Skempton (1970) has found that the mean water content in the top 25 cm of a sea bed mud is 1·75 times the liquid limit. At depth the moisture content is normally between the plastic and liquid limits. An example of the distribution of properties of a recent estuarine mud is shown in Fig. 1.

The ratio between the shear strength (c) and the overburden pressure (p) varies with water content, plastic limit and clay fraction. As these characteristics vary with the sedimentation rate, Moore (1961) has suggested that the ratio c/p may be a means of determining the sedimentation rate. The stability of slopes in mud is also governed by the variation of c and p with depth.

When the rate of deposition is high, on delta fronts for example, the pore water may have insufficient time to escape to allow complete consolidation. These deposits, therefore, will be underconsolidated. On the Mississippi delta front the sediments are underconsolidated, the rates of accumulation are 30 cm/year and would become unstable on $1\frac{1}{2}°$ slopes if their thickness exceeded 25 m (Moore, 1961).

The initial strength of the surface sediment depends on the composition of the bed material and on its moisture content. There is a minimum scouring shearing stress at which the smallest particles are eroded and carried into suspension. According to Migniot (1968) this is a function of the square root of the strength for concentrations less than about 300 g/l and a function of the strength above 300 g/l. The boundary between these two relationships may correspond to the end of the fluid mud phase. The increase in strength with decreasing moisture content reflects the increased cohesive forces caused by additional interparticle bonds. For San Francisco mud, Einstein and Krone (1962) found that during the fluid mud phase the shear strength of the mud and the fluid shears

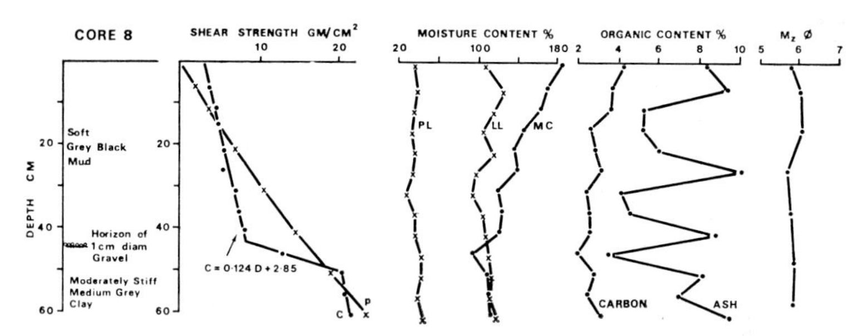

FIG. 1. Properties of a core of recent estuarine mud from Southampton Water: c shear strength, p overburden pressure, PL plastic limit, LL liquid limit, MC natural moisture content, M_z mean grain size in phi units.

that could be resisted by the sediment, rose linearly, being about 1 dyne/cm^2 at a concentration of about 20 g/l, to an extrapolated value of about 10 dyne/cm^2 at the end of the fluid mud phase.

The rate at which the bed is eroded depends mainly on the shear stress imposed by the flow on the bottom. Partheniades (1965), also working with San Francisco mud, found a minimum scouring stress of about 0·6 dyne/cm^2 and a rate of erosion that increased rapidly with increasing bed shear after a critical shear stress of between 4·8–13·4 dyne/cm^2. Peirce *et al.* (1970), in experiments on a variety of muds, have shown similar results with general erosion of the bed between 16–160 dyne/cm^2. Allen (1969) has examined the variation of bed feature form with fluid shear and found that longitudinal grooves occurred at low flows. Similar features have been described by Dyer (1970) in Southampton Water.

Consequently it appears that during a tidal cycle as the current diminishes individual particles can settle and adhere to the bed, or if the concentration is sufficiently high, fluid mud can form. The mud will gradually consolidate during the slack water period and, as the current increases at the next stage of the tide, erosion may not be intense enough to remove all of the material deposited. Deposition and re-erosion must be a continuous sequence of events as generally there is more sediment in suspension than is required to complete a year's sedimentation. Sedimentation rates for estuarine muds average 0·2 cm/year (Skempton, 1970).

TRANSPORT OF SUSPENDED MATERIAL

In a unidirectional shear flow, such as may be expected in rivers, the velocity increases exponentially, and the suspended sediment concentration decreases exponentially towards the surface. Though the grains tend to fall by gravity, their fall is resisted by an exchange of momentum between the turbulence and the particles. For low concentrations this can be expressed as:

$$\varepsilon_s \frac{\partial C}{\partial z} + Cw = 0 \qquad \text{or} \qquad \frac{w}{\varepsilon_s} = -\frac{d(\ln C)}{dz}$$

where ε_s is the diffusion coefficient for sediment, normally considered to be the same as that for fluid. C is the sediment concentration and w the sediment fall velocity. Individual particles are not necessarily maintained in suspension as there is a continuous exchange between the suspension and the bed; particles reach the bed only to be re-suspended.

The concentration at a height z relative to that at a height a can be expressed as:

$$\frac{C_z}{C_a} = \left(\frac{h - z}{h - a} \cdot \frac{a}{z}\right)^{w/kU_*}$$

where U_* is the friction velocity ($U_* = \sqrt{\tau_0/\rho}$, where τ_0 is the bed shear stress) and k is von Karman's constant ($k = 0{\cdot}4$ normally, but under high suspended sediment concentrations it may be reduced as low as 0·2). The concentration consequently rises with the current strength. The suspended load transport is the depth integral of the product of velocity and concentration.

Bagnold (1966) considers that under certain conditions the transport rate i_s should vary with the stream power according to $i_s = 0{\cdot}01(W\bar{u}/w)$ where W, the stream power, is $\tau_0\bar{u}$, $\bar{u}$ being the mean velocity.

In rivers the suspended sediment concentration C (mg/l) varies with the discharge Q (m^3/s) according to $C = aQ^b$ where a ranges between 0·004 and 80 000 and the exponent b varies between 0·0 and 2·5 (Müller and Förstner, 1968). Figure 2 shows the variation of suspended load discharge during a year for the Clyde.

In the lower parts of the Humber and the Mersey, however, the mean suspended sediment concentration is determined more by temperature and tidal range than by river flow (Jackson, 1964; Halliwell and O'Connor, 1966). In the Mersey Narrows the landward transport of suspended sediment rises rapidly with increasing tidal range, though the suspended sediment concentration and transport is generally higher on tides decreasing from spring tide stage than on those increasing from neaps.

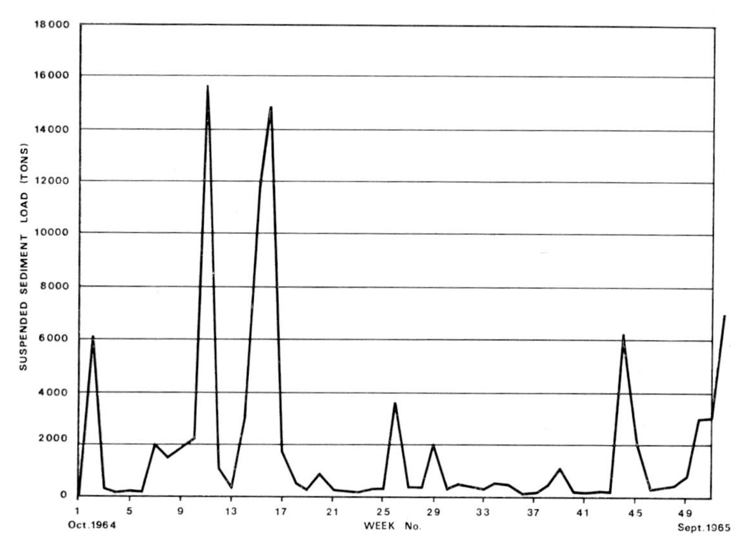

FIG. 2. Suspended sediment load discharge curve for the River Clyde at Daldowie for 1964–1965. From Fleming (1970).

In the tidal reaches of the river, below the boundary of the horizontal tide, the water flow can be reversed for part of the tidal cycle though the net flow is still seaward. Figure 3 shows the velocity and suspended sediment concentrations for a station in the upper part of Chesapeake Bay during the spring high river discharge. The maximum concentration is near the bottom and due to lag effects occurs from one to two hours after the maximum current. There is only a slight peak associated with the flood current. Around slack water the concentration is about even from surface to bottom. The net sediment transport is downstream.

In the stratified parts of the estuary the distribution of suspended matter is more complex. The stratification can be destroyed by the intensity of the mixing at times of maximum flow, especially at constrictions. Peak concentrations are now associated with both flood and ebb currents, though still with a lag (Fig. 4). At the surface the maximum velocity occurs during the ebb and on the bottom

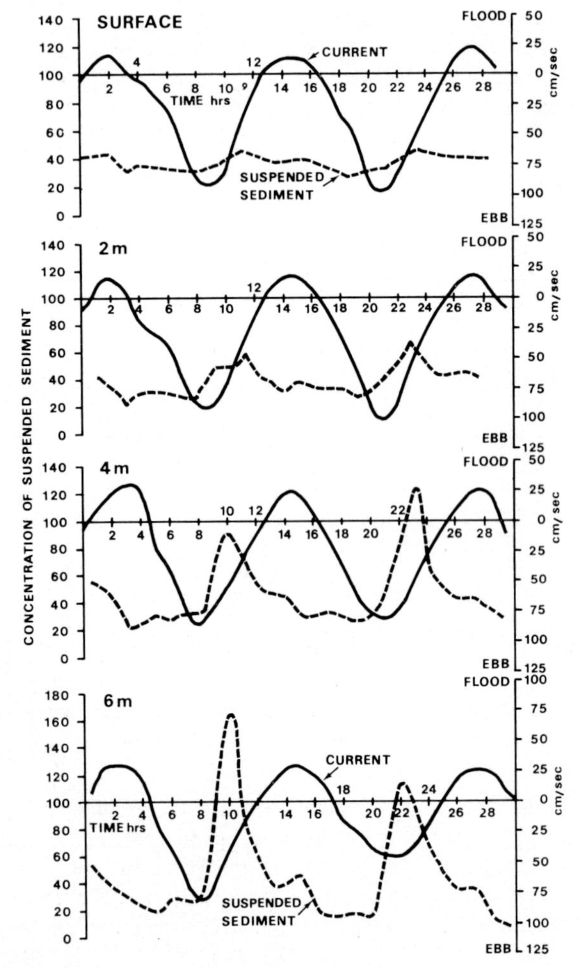

FIG. 3. Velocity and suspended sediment concentrations at Stn B9 Upper Chesapeake Bay at high river discharge. From Schubel (1969a).

during the flood. There is a net seaward suspended sediment transport near the surface and a net landward transport near the bottom. This process is further illustrated in Fig. 5.

Sheldon (1968) has shown that the high concentrations in the

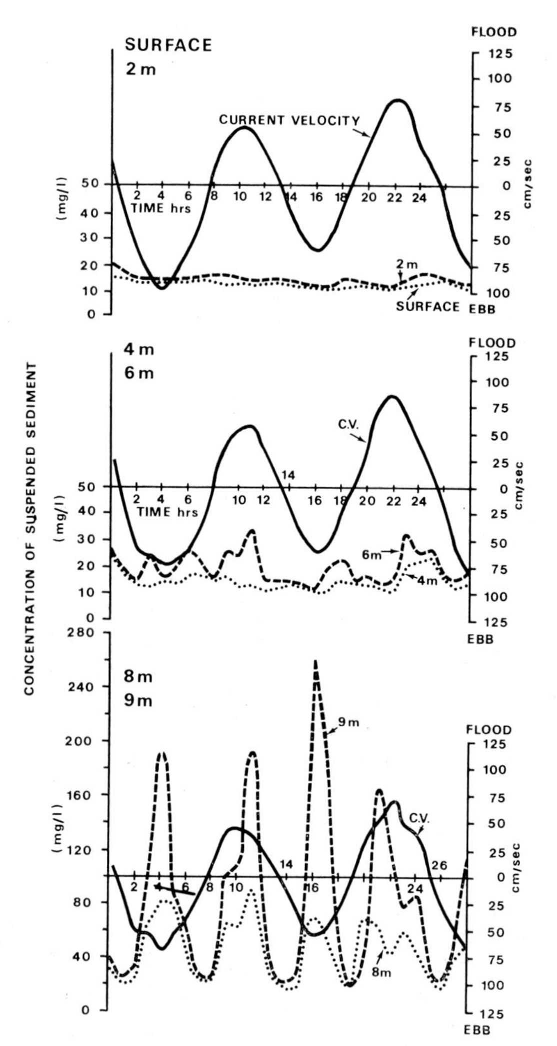

FIG. 4. Velocity and suspended sediment concentrations near Stn IIIC Upper Chesapeake Bay at low river discharge. From Schubel (1969a).

Crouch estuary associated with the flood tide were maintained for about three hours. On the ebb tide the maximum was shorter lived. In the Wadden Sea, Postma (1961) found that there was a long time at high water with low currents because of tidal asymmetry and the sediment settled out. At low tide most of the material remained in suspension because of a short slack water period.

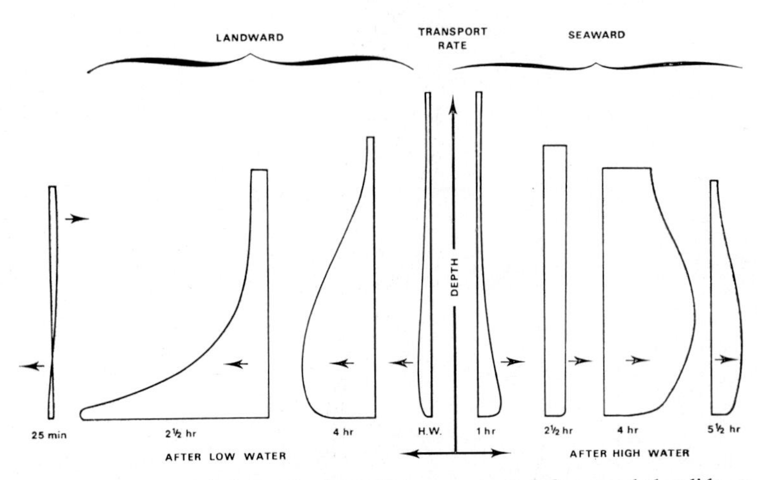

FIG. 5. Typical vertical distributions of transport rate of suspended solids at various stages of the tide, 19 miles below London Bridge, River Thames. From Inglis and Allen (1957). (Reproduced by permission of the Council of the Institution of Civil Engineers.)

In many estuaries there seems to be a natural background of 10–20 mg/l of suspended sediment, of an equivalent diameter of 3–4 microns, present at all stages of the tide. These particles may be finely divided organic matter, living planktonic creatures and fine clay particles, with fall velocities of the order of 10^{-3} cm/s. These fall velocities are of the same magnitude as the mean velocity of vertical secondary currents and the motion associated with vertical diffusion. As the particles would fall less than a metre a day in still water they must be almost perpetually in suspension.

During the tidal cycle there may be changes in the particle size distribution of the suspended sediment. The coarsest particles are nearest the bottom, but Postma (1961) and Sheldon (1968) report little variation in the particle size distribution during the tidal cycle. This implies that increasing currents do not preferentially pick up more of one size than another. Schubel (1971), however, reports a size distribution displaced to larger sizes with increasing velocity. The size distribution near the bottom was positively skewed at all times, but skewness decreased near times of maximum current as more large particles were re-suspended.

Greater differences in size distributions are apparent in the lateral and longitudinal directions. Price and Kendrick (1963) report large variations in the proportions of sand present in suspension across the Mersey Narrows. Most estuaries show a decrease in grain size towards the estuary head (Postma, 1961; Nichols and Poor, 1967) and an increase in the mean concentration of suspended matter. Higher concentrations are reached in the upper estuary than in the river or in the lower estuary. This 'turbidity maximum' is sustained by the mean estuarine circulation forming a semi-closed system. Particles entering from the river on the mean seaward surface flow settle into the bottom landward flowing layer in the lower estuary and, joining particles entering from the sea, are carried back to the upper estuary. There the sediment is mixed into the surface layer by the tidal action and a continuous exchange of particles takes place. This action forms a very effective sorting mechanism and the size range of particles

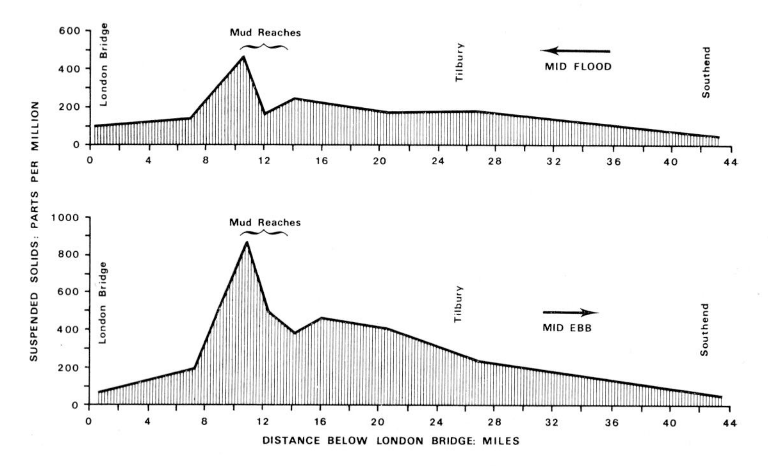

FIG. 6. Typical longitudinal distributions of suspended solids at mid-flood and mid-ebb at 0·6 of the water depth, River Thames. From Inglis and Allen (1957). (Reproduced by permission of the Council of the Institution of Civil Engineers.)

involved in the maximum is narrow (Schubel, 1969b), the coarser particles being deposited and the finer ones being swept through. The total amount of sediment involved in the turbidity maximum must be considerable. In the Thames (Fig. 6) the maximum coincides with the 'mud reaches'. Increased tidal range and river flow pushes this maximum seawards by up to twelve miles, but these shifts are of too short a duration to have a permanent effect on the sediment distribution (Inglis and Allen, 1957). The turbidity maximum will be located near the head of the salt intrusion, but diffusion can carry suspended particles farther upstream against the mean flow. This is shown by the presence of marine ostracods up to fourteen miles from London Bridge in a normally oligohaline zone (Prentice *et al.*, 1968). The turbidity maximum for the Rappahannock (Fig. 7) shows the greatest concentrations occurring near the bottom at the head of the estuary. The cross-section of the suspended load transport rates shows seaward movement occurring in the shallows on the estuary sides and landward movement concentrated in the deep channel. Lower down this estuary, Coriolis force causes the downstream transport to predominate on the right-hand side.

A qualitative model describing the landward transport of sediment, particularly in shallow tidal flat areas, has been developed by van Straaten and Kuenen (1958) and Postma (1961). If the maximum current velocity decreases towards the shore, particles drifting with

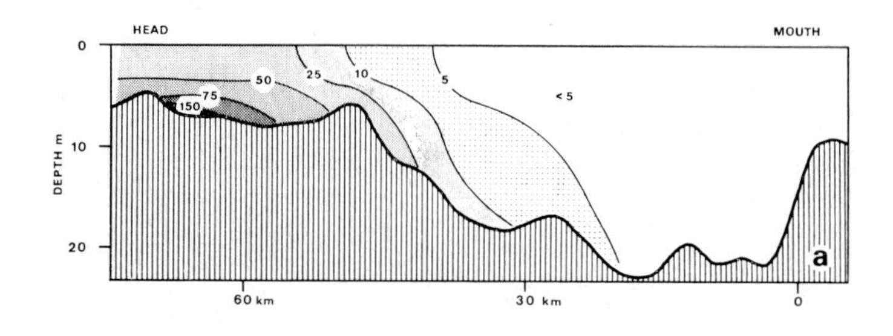

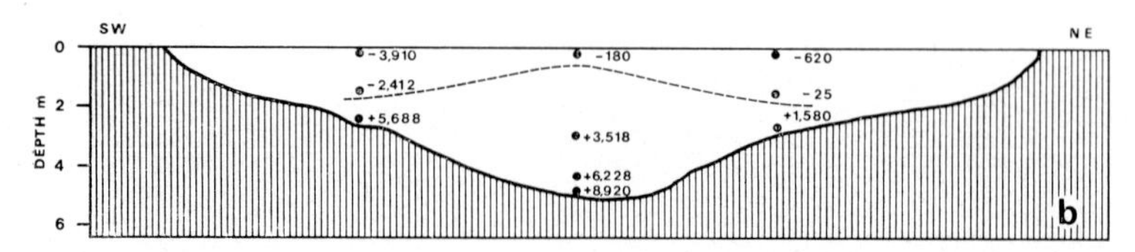

FIG. 7. Rappahannock Estuary. (a) Longitudinal distribution of average total suspended sediment concentration in mg/l. (b) Transport rate of total suspended sediment in cross-section in middle estuary. Dashed line, level of no net motion. Positive values upstream transport. Negative, downstream transport. View headward. Rates in g/m^2/hour. From Nichols and Poor (1967).

the current will undergo a changing velocity with distance along the channel during the tide. The exact form of the distance-velocity curve will depend on the asymmetry of the tidal curve, caused by the faster progression of the tidal wave in deeper water. The total envelope of the distance-velocity curves will be the shorewards decreasing maximum current. The concept of settling lag and scour lag are now introduced. Settling lag is the time taken for a particle of sediment to reach the bottom when the decreasing current cannot maintain it in suspension. Scour lag is the delay due to the difference in the current necessary to keep a particle in suspension and that required to lift it into suspension. At about high water, as there is a long period of nearly slack water, most of the particles have time to settle to the bottom. Because of settling lag the point where they reach the bottom will be landward of that at which they started settling. During the ebb tide, because of scour lag, the particle will eventually be re-suspended by water that started landward of that which deposited the particle on the flood. Consequently around high water the particles undergo an inward translation of the locus of their movement and at low water, due to the short period of slack water, no corresponding outward shift occurs. These effects will be enhanced by the average high water depth being less than the average depth at low water. Additionally at high water intertidal plants and animals trap the sediment.

The greatest differences between high and low tide occur in water masses moving in turn in a tidal channel and over a tidal flat. Thus the most efficient accumulation occurs at the inner ends of the tidal channels and on the outer parts of the flats, though the tidal mudflats can only build up to a level consistent with the effect the steeper slopes may have in increasing current velocities and subsequent scouring. The sediment moves so much before final deposition that the material is well sorted. Combined with wave action, these processes can produce a zonation of the shore such as that described by Evans (1965) and Haynes and Dobson (1969).

An excellent description of the current flow patterns over tidal flats and channels is given in Erskine Childers' novel *The Riddle of the Sands*.

TRANSPORT OF BEDLOAD MATERIAL

It is difficult to distinguish bedload and suspended load in an estuary since what is bedload at one stage of the tidal cycle may be suspended at another time. However, the bedload material is normally composed of the larger, sand size material, predominantly quartz. These grains have low surface charges and act as separate, cohesionless particles. Their characteristics have been more fully investigated by flume studies than cohesive sediments and the results are more consistent and explicable.

The threshold of movement of cohesionless grains has been studied over a wide range of grain sizes. The bed shear stress can be written in dimensionless terms as $\theta = \tau_0/(\sigma - \rho)gD$ where τ_0 is the bed shear stress, σ and ρ are the grain and fluid densities and D the grain diameter. It has been found over a wide range of sizes greater than about 200 microns, that the grains move when θ exceeds about 0·06. Below this size the threshold value of θ increases to about 0·3 at 10 microns (White, 1970).

Bagnold (1966) has introduced a suspension criterion that separates the realms of bedload sand movement and suspension. Suspension occurs when $\theta > KW^2/gD$, where K is a constant with a value of 0·4 (Bagnold, 1966) or 0·19 (McCave, 1971). In the bedload regime $0{\cdot}06 < \theta < 0{\cdot}4$ the sediment surface is rippled or duned. Ripples are features whose scale depends on the grain size and dunes are those whose scale is related to the depth of flow. Generally dunes have heights between 0·2 and 0·1 of the water depth.

Dunes are often called sand waves or megaripples and usually have an asymmetrical profile which indicates their direction of movement. Allen (1963) quotes observations in the River Loire that give the velocity of ripple movement as $Vr = 2{\cdot}65 \times 10^{-7}\ V^{2{\cdot}41}$, where V is the stream velocity in cm/s. Salsman *et al.* (1966) have measured the rate of advance of sand waves under tidal conditions in Florida as averaging 1·35 cm/day. Movement only occurred during flood tides and was maximum at the solstices. The sand wave velocity V_w (cm/day) was related to the mean flood flow velocity V (cm/s) by $V_w = 3{\cdot}12 \times 10^{-7}\ V^5$. According to Farrell (1970) intertidal sand waves moved only at spring tides and averaged 3·5 feet per tidal cycle, equivalent to 90 feet per year.

According to Bagnold (1966) the bedload transport rate at the high flow stage is proportional to the power of the flow. At lower flows the constant of proportionality, determined by Kachel and Sternberg (1971) from observations on the movement of ripples, was

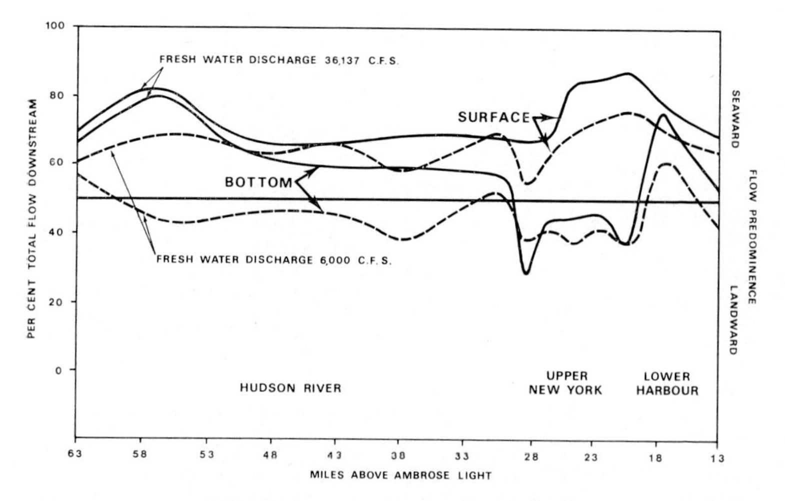

FIG. 8. Flow predominance, surface and bottom for high and low river discharges, Hudson River. From Duke (1961).

still related to the excess bed shear stress. These and other analyses suggest that the transport rate is proportional to the bed shear stress to the power 3/2 to 3. Consequently in oscillatory flow the direction and rate of long-term bedload transport will be governed by small inequalities in the flow strength.

In the river-dominated reaches, bedload material will move downstream to the top of the saline intrusion. Material entering from the sea will tend to be carried landwards until the currents are unable to move particular sizes. Thus in an ideal partially-mixed estuary there is a decrease in grain size landwards. This is upset by local constrictions which will create coarser patches. Widening, by reducing current velocities, tends to produce localised areas of deposition. Duke (1961) has shown that even at low river flows, landward movement of bottom water in the Hudson occurs in the lower part of the estuary, seaward of an abrupt widening (Fig. 8). At that point the estuary changes from well mixed to partially mixed and major shoaling occurs.

Ideally in salt wedge estuaries the sediments should become finer towards the sea as there is little bedload transport in the salt wedge. Computer modelling of salt wedge estuary sedimentation bed has been carried out by Farmer (1971). In most fjords sedimentation is also mainly confined to the head and in many cases this leads to slope failure (Terzaghi, 1957) and slumping into the deeper areas.

SECONDARY CURRENTS

In rivers there is usually a meandering flow with scour holes on the bends. Within the meanders a secondary helical flow system is set up directed downwards into the scour holes and upwards in the shallows on the inside of the bend. At times of peak flow the near bottom velocity in the scour hole is greater than that on the riffle and at low flow stages the reverse holds (Keller, 1971). Consequently finer material selectively deposited in the scour holes at low flow stages will be flushed out during floods. Coarser material should be present on the riffles. If a saline intrusion occurs when river flow diminishes, the scour holes can contain stagnant pools of saline water over most of the tidal cycle (Dyer and Ramamoorthy, 1969) with suspended sediment deposited preferentially in the scour holes rather than on the riffles. In the Vellar estuary the sediment distribution (Fig. 9) shows finer, negatively skewed sand in the scour holes and coarser positively skewed sand on the riffles.

In regions dominated by tidal movement, there can be marked differences in the strength of ebb and flood currents at different

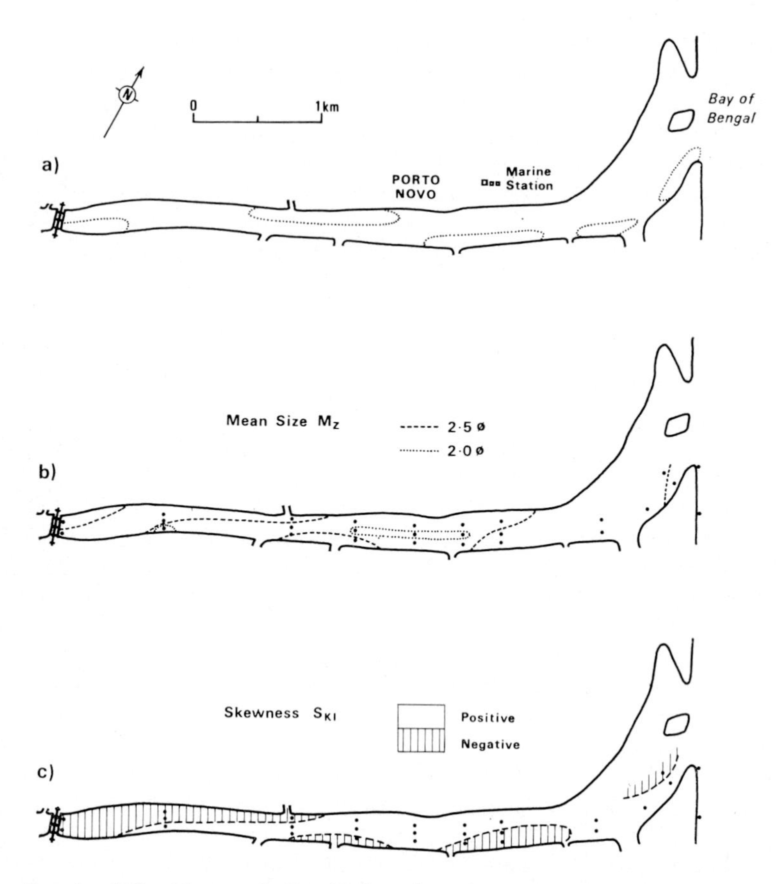

FIG. 9. Vellar Estuary, India; (a) Location of scour holes in lower estuary; (b) Distribution of mean grain size M_Z in phi units. Dots, sample positions; (c) Distribution of inclusive graphic skewness S_{KI}.

positions in the channel. These variations produce horizontal secondary current systems and may be partly due to the effect of Coriolis force concentrating flows in different parts of the channel. Inglis and Allen (1957) have shown that the net suspended load transport nineteen miles below London Bridge is landward on the northern side and seaward on the southern. This situation is generally associated with meanders in the channel and is caused by the flood currents taking a straighter course than the ebb currents which tend to follow the meanders. In the Taw-Torridge estuary, in contrast, the reverse situation occurs (Prentice *et al.*, 1968) possibly due to the large area of banks and shallows. The drying and covering of banks creates complicated water flows and patterns of sediment dispersal. Klein (1970) describes the effects of this situation in the Bay of Fundy.

In the broader part of the estuary the flows may become distinctly separated into 'ebb and flood channels'. The former contain predominant ebb currents and shallow and narrow towards the sea. The latter narrow and shallow in a landward direction and contain predominant flood currents. Quite often these channels occur in pairs, one ending at a steep slope on the edge of the other. As the tidal excursion is generally longer than a channel's length, the water flows up one channel and down the other, forming a circulating system which the sediment also tends to follow. The channels shift their position in an apparently consistent manner (Inglis and Kestner, 1958) and the banks between vary in extent and volume (Cloet, 1967). These movements of the channels cause fluctuation of as much as 5% in estuary volume and are the process by which accumulation is kept in check and progressive deterioration prevented (Inglis and Kestner, 1958). The circulations mainly involve bedload material and the continual reworking provides a good means of size sorting the sediment. The penetration of bedload material into the estuary is also effectively restricted by the circulatory movement.

Near estuary mouths wave action and littoral drift also play a great part in determining the configuration of banks and channels and the supply and distribution of sediment (*e.g.* Allen, 1971). The contribution of beach or off-shore material and its distribution in the lower estuary can often be assessed by consideration of heavy mineral distributions (*e.g.* Byrne and Kulm, 1967; Windom *et al.*, 1971).

SEDIMENT BALANCE IN ESTUARIES

The relative importances of the different sediment sources in different estuaries can be quantified by drawing up a sediment balance. There has been a fall of about 10% during the last 100 years in the volume of the Mersey estuary despite the dredging of over 400 million cubic yards of material. As the volumes of incoming solid sewage matter and of material suspended in the river flow were not a major source, the material must have moved in from the sea (Price and Kendrick, 1963).

In the Clyde, however, little material is derived from the sea (Fleming, 1970). The rivers discharged 203 758 tons of suspended load and 9 189 tons of bedload annually. Sewage discharge and spillage from harbours and docks accounted for a further 29 829 tons.

These sources totalled 242 776 tons a year. Dredging removed 317 769 tons and as 75 000 tons of this was capital dredging, the sediment entering from the rivers and depositing in twelve miles of the upper reaches was balanced by dredging.

Biggs (1970) has examined the sources and distribution of suspended sediment in the inner part of Chesapeake Bay. As can be seen from Fig. 10, the river contributes over 80% of the suspended

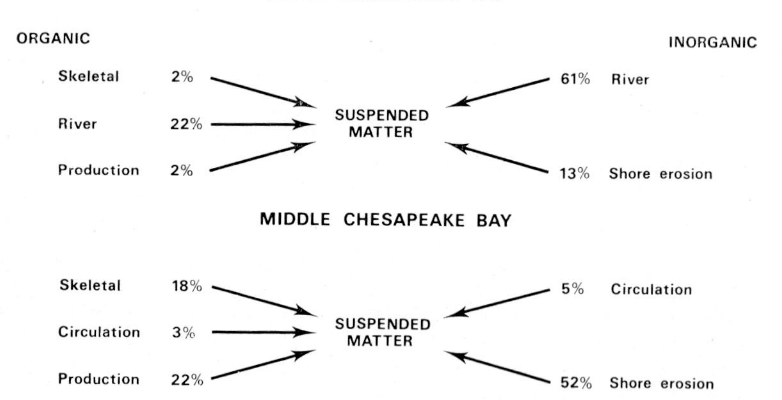

FIG. 10. Contributions to total suspended sediment load by organic and inorganic sources in the Upper and Middle Chesapeake Bay. From Biggs (1970).

material in the upper bay. Lower down the organic contribution is larger, but shore erosion becomes the main sediment source. Of the suspended sediment, 4% of the upper bay load escaped to the middle bay and 23% of the middle bay load escaped to the lower bay. The net amounts sedimented out in the upper and middle bays, if evenly distributed, would have amounted to 3·7 and 1·1 mm/year respectively.

THE INFLUENCE OF MAN

Attempts by man to stabilise the sedimentation patterns in estuaries and to alter their topography lead to disturbance of the estuaries' natural balance. Robinson (1963) has discussed the navigational consequences of estuarine sedimentation and dredging on English ports.

Creation of new docks and reclamation of marshes decreases the tidal prism, whereas dredging increases the estuary volume. This will alter flushing rates, the intensity of vertical mixing and density current effects. A model study of dredging in the James River (Nichols, 1968) showed slight freshening of the near surface waters, mainly over shoals, and increased salinity near the bottom. Vertical mixing was reduced and there was a trend for a reduction of transport in both upper and lower layers.

Dredging of the channel in the Thames increased the rate of progression of the tide and increased the tidal range, leading to increased sedimentation in the upper reaches. The tidal range above Old London Bridge was also increased by 25% by removal of the bridge (Inglis and Allen, 1957).

It is considered that the Mersey was in equilibrium at the beginning of the century and the fall in capacity, previously mentioned, is due to the training works in Liverpool Bay (Price and Kendrick, 1963). The training walls have concentrated the ebb stream and areas in the Bay which were previously ebb dominated have become flood dominated. This increased the amount of sand available to be carried landwards on the density circulation once the walls were overtopped.

Deforestation leads to increased sediment discharge and though control of river flow by damming the rivers may remove much of the sediment, shoaling problems may be aggravated. This process becomes apparent by reference to Fig. 8. At high discharge the upper part of the Hudson River estuary has a net seaward movement on the bottom, but at low flows landward bottom movement occurs. Control of the river discharge at a low level would promote shoaling at the head of the estuary as well as that at present occurring lower down at all river flows.

The effect of altering river flows is shown also by the diversion of the Santee River into Charleston Harbour. The drainage area and sedimentation were both increased by an order of magnitude with a similar increase in dredging costs. The annual sediment supply rose to 3·5–6 million cubic yards and examination of heavy and clay minerals showed that this came from the new drainage area (Neiheisel and Weaver, 1967).

Where meandering within the estuary is suppressed there is a resulting loss of volume (Price and Kendrick, 1963). In the Lune, though the immediate effect of training the meanders was to improve the channel depth, eventually deposition outside the channel halved the cubature and the channel depth reverted to its pre-training value (Inglis and Kestner, 1958). As the solution to a shoaling problem in the Mersey, however, it was suggested that reduced dredging might eventually lead to naturally increased depths (Price and Kendrick, 1963).

Consequently, sedimentation in estuaries appears to be limited by a sort of servo-control mechanism. Deposition causes alteration of the current flow and this alteration tends to limit the amount of deposition that can occur. Thus an uneasy natural balance is maintained. Alteration of the system beyond the natural limits will cause changes that, in many ways, are still unpredictable.

ACKNOWLEDGEMENTS

I wish to thank the authors and journals concerned for permission to reproduce figures from their publications.

REFERENCES

Allen, G. P. (1971). 'Relationship between grain size parameter distribution and current patterns in the Gironde Estuary (France).' *J. Sed. Petrology*, **41,** 74–88.

Allen, J. R. L. (1963). 'Asymmetrical ripple marks and the origin of water laid cosets of cross-strata.' *Geol. Jour.*, **3,** 187–236.

Allen, J. R. L. (1969). 'Erosional current marks of weakly cohesive mud beds.' *J. Sed. Petrology*, **39,** 607–23.

Allersma, E., Hoekstra, A. J. and Bijker, E. W. (1966). 'Transport patterns in the Chao Phya estuary.' 10*th Conf. Coastal Engin.*, **1,** 632–50.

Bagnold, R. A. (1966). 'An approach to the sediment transport problem from general physics.' *U.S. Geol. Surv. Prof. Paper No.* 422–1, 37 pp.

Biggs, R. B. (1970). 'Sources and distribution of suspended sediment in Northern Chesapeake Bay.' *Marine Geol.*, **9,** 187–201.

Byrne, J. V. and Kulm, L. D. (1967). 'Natural indicators of estuarine sediment movement.' *Proc. Amer. Soc. Civil Eng.*, **93,** WW2, 181–94.

Cloet, R. L. (1967). 'Marine sediment circulations.' *Conf. Tech. Sea and the Sea-bed*, 546–63.

Duke, C. M. (1961). 'Shoaling of the lower Hudson River.' *Proc. Amer. Soc. Civil. Eng.*, **87,** WW1, 29–45.

Dyer, K. R. (1970). 'Linear erosional furrows in Southampton Water.' *Nature, Lond.*, **225,** 56–8.

Dyer, K. R. and Ramamoorthy, K. (1969). 'Salinity and water circulation in the Veller estuary.' *Limnol. Oceanog.*, **14,** 4–15.

Einstein, H. A. and Krone, R. B. (1962). 'Experiments to determine modes of cohesive sediment transport in water.' *J. Geop. Res.*, **67,** 1451–61.

Evans, G. (1965). 'Intertidal flat sediments and their environment of deposition in the Wash.' *Quart. J. Geol. Soc. Lond.*, **121,** 209–41.

Farmer, D. G. (1971). 'A computer simulation model of sedimentation in a salt wedge estuary.' *Marine Geol.*, **10,** 133–43.

Farrell, S. C. (1970). 'Sediment distribution and hydrodynamics Saco River, and Scarboro estuaries, Maine.' Cont. 6-CRG. *Dept. Geol. Univ. Mass.*

Fleming, G. (1970). 'Sediment balance of Clyde Estuary.' *Proc. Amer. Soc. Civil Eng.*, **96,** H.Y.11, 2219–30.

Guilcher, A. (1963). 'Estuaries, deltas, shelf, slope.' In *The Sea*, Vol. 3. Ed. M. N. Hill. New York: Wiley.

Halliwell, A. R. and O'Connor, B. A. (1966). 'Suspended sediment in a tidal estuary.' 10*th Conf. Coastal Eng.*, **1,** 687–706.

Hansen, D. V. and Rattray, M. Jr (1966). 'New dimensions in estuary classification.' *Limnol. Oceanog.*, **11,** 319–26.

Haynes, J. and Dobson, M. R. (1969). 'Physiology, foraminifera and sedimentation in the Dovey estuary (Wales).' *Geol. J.*, **6,** 217–56.

Inglis, Sir C. C. and Allen, F. H. (1957). 'The regimen of the Thames estuary as affected by currents, salinities and river flow.' *Proc. Inst. Civil Eng.*, **7,** 827–68.

Inglis, Sir C. C. and Kestner, F. J. T. (1958). 'The long term effects of training walls, reclamation and dredging on estuaries.' *Proc. Inst. Civil Eng.*, **9,** 193–216.

Jackson, W. H. (1964). 'Effect of tidal range, temperature and freshwater on the amount of silt in suspension in an estuary.' *Nature, Lond.*, **201,** 1017.

Kachel, N. B. and Sternberg, R. W. (1971). 'Transport of bedload as ripples during an ebb current.' *Marine Geol.*, **10,** 229–44.

Keller, E. A. (1971). 'Areal sorting of bed-load material: The hypothesis of velocity reversal.' *Geol. Soc. Amer. Bull.*, **82,** 753–6.

Klein, G. de V. (1970). 'Depositional and dispersal dynamics of intertidal sand bars.' *J. Sed. Petrology*, **40,** 1095–1127.

Lauff, G. H. (Ed.) (1967). *Estuaries.* Amer. Assoc. Adv. Sci., 757 pp.

McCave, I. N. (1971). 'Sand waves in the North Sea off the coast of Holland.' *Marine Geol.*, **10,** 199–225.

Meade, R. H. (1969). 'Landward transport of bottom sediments in estuaries of the Atlantic Coastal Plain.' *J. Sed. Petrology*, **39,** 222–34.

Migniot, C. (1968). 'Study of the physical properties of various forms of very fine sediments and their behaviour under hydrodynamic action.' *La Houille Blanche*, **23,** 591–620.

Moore, D. G. (1961). 'Submarine slumps.' *J. Sed. Petrology*, **31,** 343–57.

Müller, G. and Förstner, U. (1968). 'General relationship between suspended sediment concentration and water discharge in the Alpenrhein and some other rivers.' *Nature, Lond.*, **217,** 244–5.

Neiheisel, J. and Weaver, C. E. (1967). 'Transport and deposition of clay minerals, Southeastern United States.' *J. Sed. Petrology*, **37,** 1084–1116.

Nichols, M. M. (1968). 'Effect of channel deepening on salinity in the James Estuary.' 11*th Conf. Coastal Eng.*, 1439–41.

Nichols, M. M. and Poor, G. (1967). 'Sediment transport in a coastal plain estuary.' *Proc. Amer. Soc. Civil Eng.*, **93,** WW4, 83–95.

Partheniades, E. (1965). 'Erosion and deposition of cohesive soils.' *Proc. Amer. Soc. Civil Eng.*, **91,** HY1, 105–39.

Peirce, T. J., Jarman, R. T. and de Turville, C. M. (1970). 'An experimental study of silt scouring.' *Proc. Inst. Civil Eng.*, **45,** 231–43.

Peirce, T. J. and Williams, D. J. (1966). 'Experiments on certain aspects of sedimentation of estuarine muds.' *Proc. Inst. Civil Eng.*, **34,** 391–402.

Postma, H. (1961). 'Transport and accumulation of suspended matter in the Dutch Wadden Sea.' *Neth. Jour. Sea Res.*, **1,** 148–90.

Prentice, J. E., Beg, I. R., Colleypriest, C., Kirby, R., Sutcliffe, P. J. C., Dobson, M. R., D'Olier, B., Elvines, M. F., Kilenyi, T. I., Mandrell, R. J. and Phinn, T. R. (1968). 'Sediment transport in estuarine areas.' *Nature, Lond.*, **218,** 1207–10.

Price, W. A. and Kendrick, M. P. (1963). 'Field and model investigation into the reasons for siltation in the Mersey estuary.' *Proc. Inst. Civil Eng.*, **24,** 473–517.

Pritchard, D. W. (1952). 'Estuarine hydrography.' *Advances Geop.*, **1,** 243–80.

Pritchard, D. W. (1954). 'A study of the salt balance in a coastal plain estuary.' *J. Marine Res.*, **13,** 133–44.

Pritchard, D. W. (1955). 'Estuarine circulation patterns.' *Proc. Amer. Soc. Civil Eng.*, **81,** 717.

Robinson, A. H. W. (1963). 'Physical factors in the maintenance of port approach channels.' *Dock Harb. Auth.*, **43.**

Salsman, G. G., Tolbert, W. H. and Villars, R. G. (1966). 'Sand ridge migration in St Andrew Bay, Florida.' *Marine Geol.*, **4,** 11–19.

Schubel, J. R. (1969a). 'Distribution and transport of suspended sediment in Upper Chesapeake Bay.' *Tech. Rept.* 60. *Ref.* 69-13. Chesapeake Bay Institute, John Hopkins University.

Schubel, J. R. (1969b). 'Size distributions of suspended particles of the Chesapeake Bay turbidity maximum.' *Neth. J. Sea Res.*, **4,** 283–309.

Schubel, J. R. (1971). 'Tidal variation of the size distribution of suspended sediment at a station in the Chesapeake Bay turbidity maximum.' *Neth. J. Sea Res.*, **5,** 252–66.

Sheldon, R. W. (1968). 'Sedimentation in the estuary of the River Crouch, Essex, England.' *Limnol. Oceanog.*, **13,** 72–83.

Shepard, F. P. (1948). *Submarine geology*. New York: Harper. 557 pp.

Simmons, H. B. (1955). 'Some effects of upland discharge on estuarine hydraulics.' *Proc. Amer. Soc. Civil Eng.*, **81,** 792.

Skempton, A. W. (1970). 'The consolidation of clays by gravitational compaction.' *Quart. J. Geol. Soc. Lond.*, **125,** 373–408.

Terzaghi, K. (1957). 'Varieties of submarine slope failures.' *Norway Geotech. Inst.*, No. 25, 1–16.

Van Straaten, L. M. J. U. and Kuenen, Ph. H. (1958). 'Tidal action as a cause of clay accumulation.' *J. Sed. Petrology*, **28,** 406–13.

White, S. J. (1970). 'Plane bed thresholds of fine grained sediments.' *Nature, Lond.*, **228,** 152–3.

Whitehouse, U. G., Jeffery, L. M. and Debbrecht, J. D. (1960). 'Differential settling tendencies of clay minerals in saline waters.' *Proc. 7th Conf. Clays Clay Mins.*, 1–79.

Windom, H. L., Neal, W. J. and Beck, K. C. (1971). 'Mineralogy of sediments in three Georgia estuaries.' *J. Sed. Petrology*, **41,** 497–504.

3

Chemical Processes in Estuaries

JOHN PHILLIPS

INTRODUCTION

In his summing up of an earlier meeting on estuaries, Hedgpeth (1967) offered the following rough definition: 'The estuarine ecosystem is a mixing region between sea and inland water of such shape and depth that the net resident time of suspended materials exceeds the flushing'. This general statement also serves to describe two essential features of estuarine chemistry: mixing processes and the importance of the solid/water interface.

A great deal of chemical information has been collected about estuaries, but few general principles have emerged from it. Perhaps one of the reasons for this state of affairs is the position of chemistry in environmental research. There is a tendency for much of the chemical work done in this field to be part of studies which are primarily biological or geological. I hope that this review will suggest some areas in which chemical studies, pursued without reference to conventional disciplinary boundaries, would help our understanding of the estuarine environment as a whole.

The approach is broadly geochemical, considering the flux of materials to and from an estuary and their cycling within it. Examples are taken from freshwater and marine processes as they seem appropriate.

DISSOLVED SOLIDS

Every year the rivers of Britain supply about twenty million tons of dissolved solids to the ocean. All this material passes through the estuarine zone, where it encounters the dissolved solids of sea water. An estuary is therefore a region of mixing between two aqueous solutions of very different chemical composition.

The properties of river water show great variations, both geographically and temporally, whereas those of sea water are relatively constant. This is illustrated by data on the pH and Eh of fresh waters (Fig. 1) compiled by Baas Becking *et al.* (1960). These parameters reflect the positions of many important equilibria in aquatic environments, including the carbon dioxide system and the iron and sulphur cycles. These in turn are linked to important biological processes such as photosynthesis, respiration and bacterial

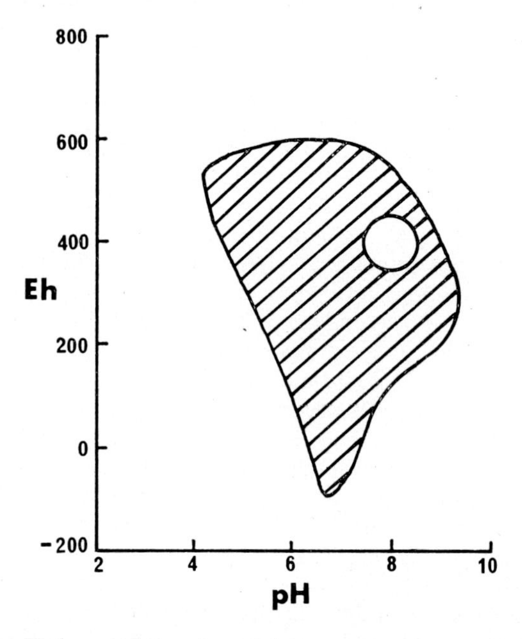

FIG. 1. Eh–pH characteristics of natural waters: cross-hatched area represents rivers and lakes, circle represents normal sea water. After Baas Becking *et al.* (1960).

activity. The narrow pH and Eh ranges of sea water are due to its greater buffer capacity and generally high degree of oxygen saturation. For the purposes of this discussion, the world average values for the composition of river water calculated by Livingstone (1963) have been employed throughout.

River waters contain, on average, about a third of one percent of the concentration of dissolved solids present in sea water. In view of this, it is tempting to regard estuarine waters simply as diluted sea water, in whose composition river water plays a negligible part. However, there are dangers in this approach, which become apparent

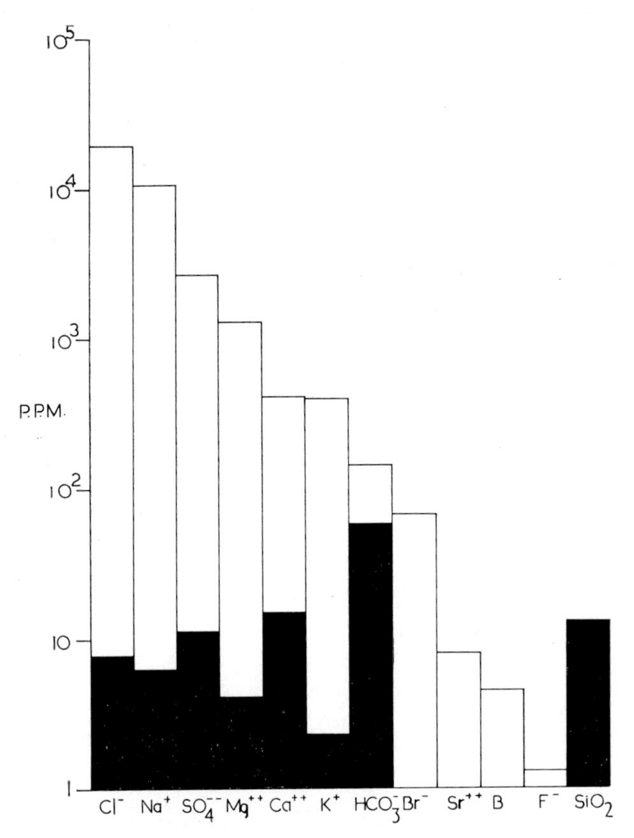

FIG. 2. Concentration of major ions in sea water at 35$^0/_{00}$ salinity (top of open bars) and world average river water (top of solid bars).

when the differences between river water and sea water are considered in more detail (Fig. 2).

The relative proportions of even the major ions of sea water, such as chloride and bicarbonate, change appreciably at low salinities because of the contribution from river water (Fig. 3). Their concentrations are of vital significance to ionic regulation and may well affect the composition of animals living in brackish waters. For example, in *Mytilus edulis* the magnesium content of the shell is inversely related to salinity (Dodd, 1965). It has been suggested that effects of this type are due to reduced discrimination between potential building materials when these are in short supply (Rucker and Valentine, 1961). A few trace metals are more abundant in sea water than in river waters. Molybdenum, for example, is usually present

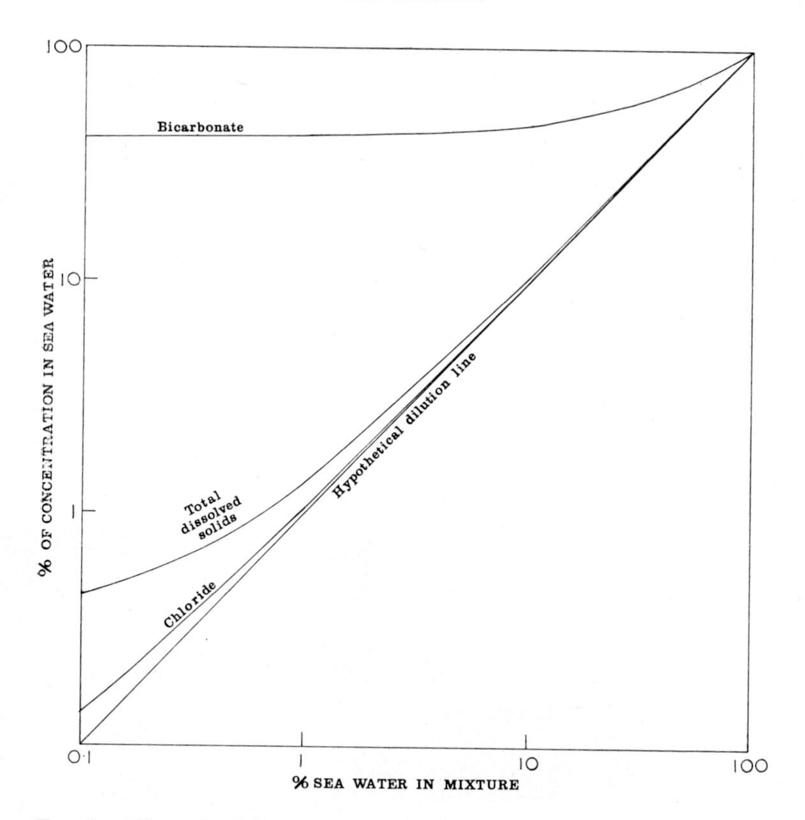

FIG. 3. Effect of mixing between sea water and river water on major ion concentrations.

in marine waters in excess of biological requirements, but molybdenum deficiency has been identified in fresh water (Goldman, 1960). The concentration of molybdenum in Southampton Water shows a positive correlation with salinity (Head and Burton, 1970).

In the case of ions which are often less abundant in coastal sea water than in river waters, such as phosphate and nitrate, their concentrations in estuaries may show a negative correlation with salinity (Fig. 4). Phosphate and nitrate are essential plant nutrients whose distribution is a significant factor in phytoplankton ecology.

Many trace elements are present at similar concentrations in both river and sea water (Durum and Haffty, 1963), but this does not imply that their behaviour is identical in both cases. For one thing activity coefficients are substantially lower in a medium of high ionic strength like sea water. This influences every chemical reaction by changing the effective concentrations of the participating chemical species. Moreover, the chemical speciation of trace elements is affected by the high concentrations of other ions present in sea water. Thus about half of the cadmium and fluorine in sea water is present as complex cations, $CdCl^+$ and MgF^+ respectively. Riley has written an excellent summary of this subject (in Riley and Chester, 1971).

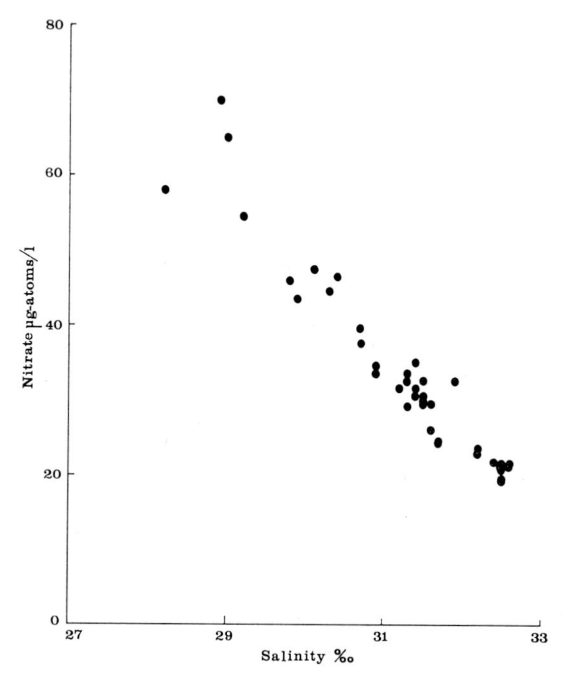

FIG. 4. Relationship between nitrate and salinity in Southampton Water on 23rd March 1970.

By coincidence, separate papers on the speciation of copper in fresh water and sea water have recently been published. Stiff (1971a) used a cupric ion-selective electrode to investigate copper-bicarbonate equilibria over the pH range of most natural waters. He added cupric ions to bicarbonate solutions and found, after several hours had elapsed, that from 75 to 99% of the copper was present as the carbonate complex $CuCO_3°$. A polarographic study of model systems led Odier and Plichon (1971) to the conclusion that dissolved inorganic copper in sea water is distributed between three main species in the proportions: 60% Cu^{++}, 37% $CuCl^+$ and 2%

$[Cu(HCO_3)_2(OH)]^-$. The formation of a chloro-complex is obviously due to the high concentration of chloride in sea water; the lack of carbonate complexing is attributable to the competition of other cations for carbonate. Therefore we must be prepared for the existence of a wide range of chemical species in estuarine water, together with the effects of varying activity coefficients.

Stiff's (1971b) analyses of river waters have revealed that the dominant fraction of dissolved copper is bound to organic matter. 'Organic' copper also accounts for a substantial fraction of the copper in sea water (Foster and Morris, 1971). This has probably been underestimated in the past because reagents for the determination of 'ionic' copper are capable of removing the element from organic complexes, so that little more is found when organic matter is destroyed.

The biological activity of trace elements is strongly influenced by speciation, which is not surprising when one considers that the properties involved are those which might be expected to affect the transport of an element across biological membranes, such as electric charge, redox status and lipophilic character. The toxicity of copper to fish is known to be reduced in hard water (Lloyd and Herbert, 1962), which suggests that carbonate complexing may be involved. Steemann Nielsen and Wium-Andersen (1970) have shown that inorganic copper is toxic to phytoplankton at levels comparable to those of total dissolved copper in natural waters. They suggest that organic complexes are responsible for reducing its toxicity.

This effect has been demonstrated in bacteria-free cultures of the diatom *Phaeodactylum tricornutum* by Spencer (1957). The toxicity of media containing the copper-EDTA complex was proportional to the concentration of free copper in equilibrium with the complex. Similarly, the response of manganese-deficient cells was proportional to the free manganese concentration and independent of the concentration of the manganese-EDTA complex. It is generally considered that the EDTA molecule is not readily metabolised.

There has been much speculation about the role of chelation in the supply of trace metals to phytoplankton. Johnston (1964) found that the addition of EDTA alone was as effective as the addition of trace metal chelates in promoting abundant growth in unfiltered sea water enriched with phosphate, nitrate and silicate. He interpreted these observations as indicating that adequate trace metals were already present in a particulate form and only required chelation to bring them into solution, where they could be collected by active sites on plant cells. He concluded that the supply of chelators is frequently the crucial aspect of phytoplankton nutrition in the sea. A suggestion that chlorophyll production in phytoplankton requires organo-iron compounds from the surrounding water has been made by Davies (1970).

Recently upwelled water in the central Pacific Ocean is high in nutrients but low in dissolved organic carbon. Barber and Ryther (1969) found that its relatively poor ability to support phytoplankton growth could be improved only by the addition of a strong chelator or a filtered zooplankton homogenate. The 'availability' of trace metal chelates is apparently no simple matter, but the wide range of

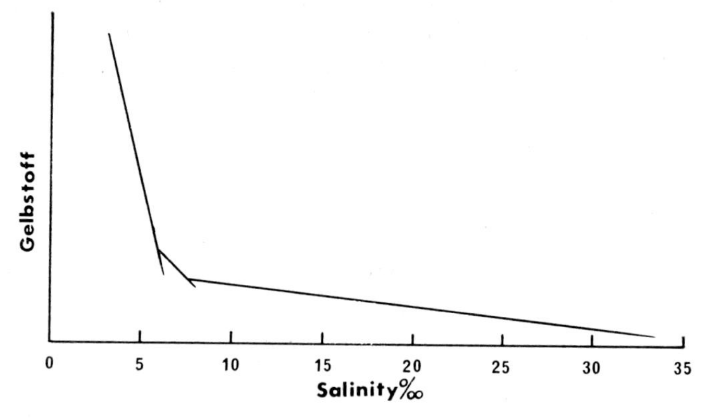

FIG. 5. Distribution of Gelbstoff relative to salinity in the Baltic Sea. After Jerlov (1955).

trace metals, organic compounds and phytoplankton species found in natural waters leaves ample room for both of the mechanisms postulated to explain the biological activity of chelators: detoxication (Steemann Nielsen and Wium-Andersen, 1970) and solubilisation (Johnston, 1964).

Fresh waters contain about ten times as much dissolved organic matter as sea water, mainly in the form of humic material derived from the leaching of soil. Similar material, the so-called Gelbstoff, is produced by marine algae. Figure 5 illustrates mixing between water masses of different salinity and humic material content. Organic compounds of this type have the ability to bind trace metals (Barsdate, 1970). In particular, the association between iron and humic acids is well known (Shapiro, 1967). The structure of humic material is complicated, but it seems to consist of large three-dimensional polyanions composed of aromatic nuclei with phenolic and carboxylic functional groups, linked together by various bridging groups, some of which contain nitrogen (Fig. 6).

Although terrigenous humic acids gradually precipitate in contact with sea water (Sieburth and Jensen, 1968), there is evidence that they have an important role in the fertility of estuaries. Sterilised soil extract was a useful addition to many of the older culture media for phytoplankton. Nelson (1947) reported a correlation between heavy rainfall, with attendant soil erosion, and blooms of the diatom *Skeletonema costatum* in the coastal waters of New Jersey. Some species of neritic phytoplankton have a higher iron requirement than oceanic forms (Ryther and Kramer, 1961). It has been demonstrated that marine dinoflagellates respond to additions of humic material, or of river water containing appreciable quantities of humic material, with improved growth rate and yield (Prakash and Rashid, 1968). Thiamine, a vitamin of the B group essential for some marine

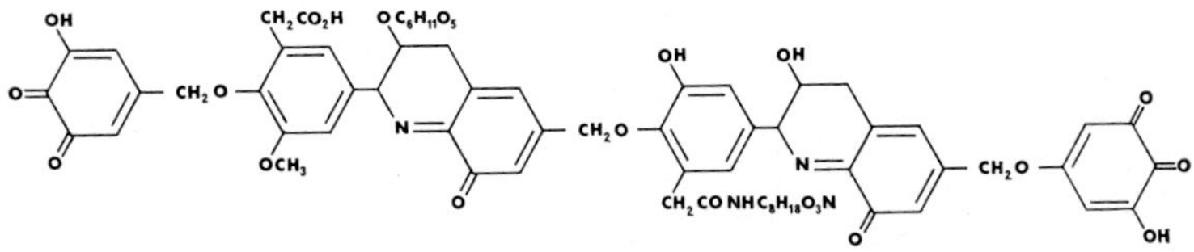

FIG. 6. A possible structure for humic acid. The fused ring system contains a potential chelation site. After Dragunov *et al.* (1948).

unicellular algae, may be derived largely from land drainage (Vishniac and Riley, 1961). The mechanisms by which the organic compounds of river water influence estuarine ecology are by no means completely understood yet.

A further source of organic material is becoming increasingly important in many estuaries. This is the discharge of sewage effluent and other wastes from human activities. At the present time about a quarter of Britain's sewage is discharged direct to coastal waters after varying degrees of treatment. Raw sewage in an estuary is an obvious form of pollution, but even a good quality effluent has its dangers. Biological treatment converts most of the phosphorus and nitrogen in sewage to inorganic forms which encourage algal blooms in the receiving waters. This is the process known as eutrophication. The organic matter resynthesised in this way eventually decomposes, exerting a biological oxygen demand on the waters of the estuary. If re-aeration is insufficiently rapid, this leads to anoxia and the destruction of a natural resource.

The phosphate content of natural waters has frequently been used as an indicator of this type of pollution, and as such it has proved very useful (Ketchum, 1969). However, this has resulted in a tendency to regard phosphate as the prime factor in eutrophication whereas, at least in coastal waters, shortage of nitrogen is most commonly the factor limiting primary production. The phosphate measured may be surplus to requirements, whilst combined inorganic nitrogen may be undetectable in eutrophic waters. Ryther and Dunstan (1971) have sounded a warning against the trend in the United States toward the replacement of phosphate in detergents with the nitrogenous substitute, nitrilotriacetic acid. As they say, this may be simply adding fuel to the fire.*

It was good news when the Royal Commission on Environmental Pollution announced in its first report that its proposals for work in the year 1971–72 are to 'enquire into and report on the problems of pollution in tidal waters, estuaries and the seas around our coasts' and 'to keep under review action to improve Britain's rivers'.

We have seen that it is possible to classify dissolved substances roughly into three groups according to their relative concentrations in river waters and in sea water: those which are more abundant in sea water (the major ions), those which are more abundant in river waters (plant nutrients, organics) and those which have similar concentrations in both (most trace metals). When we come to particulate substances, the situation seems rather simpler at first, because in general the greater velocities found in rivers maintain higher concentrations of material in suspension, whereas the comparatively slow currents found at sea favour sedimentation. Estuaries are frequently regions of net deposition, but there is a wide range of estuarine processes which interconvert dissolved and particulate matter. The emphasis shifts from mixing toward interfaces.

PARTICULATE MATTER AND INTERFACES

The silt load of a river is subject to even greater variation than its dissolved solids content, but on average it is about three times larger. In practice, the distinction between the dissolved and particulate fractions is an arbitrary one, usually made on the basis of filtration. Experiments by Sheldon and Sutcliffe (1969) have shown that membrane filters of 5 μm nominal pore size retain over 50% of particles below 1 μm equivalent spherical diameter. Some particles in the colloidal size range (approximately 1 nm–1 μm) are probably included in the dissolved fraction by such a process. Colloidal particles are stabilised in fresh water by repulsion between their electric charges, but in sea water these charges are neutralised, causing flocculation (Fig. 7). The process is reversible. Aggregation

* The use of nitrilotriacetic acid in detergents is currently banned in the United States because of unresolved questions concerning its possible long-term effects on health (*Envir. Sci. Technol.*, **5**, 747, 983 (1971)).

of dissolved organic matter in the presence of bacteria is a further source of particles (Sheldon *et al.*, 1967).

Krauskopf (1956) made a study of three processes which might control trace metal concentrations in sea water by incorporating them into the sediments: precipitation, adsorption and biological uptake. He calculated the maximum concentrations which a number of trace metals should attain from available data on the solubilities of their compounds with the anions present in sea water. He also carried out experiments in which a salt of each metal was added to

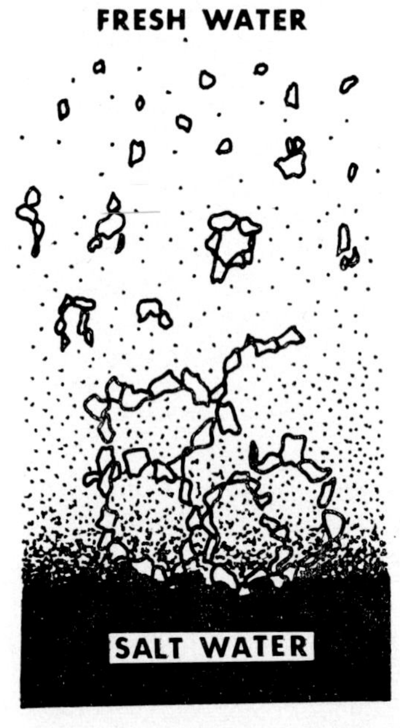

FIG. 7. Diagrammatic representation of the flocculation process. After Sakamoto (1968).

sea water until a precipitate formed, and the concentration of the metal remaining in solution was measured. The precipitate was then added to a fresh aliquot of sea water and the concentration measured again after equilibration. In this way he found upper and lower limits for the maximum concentration which each metal could attain. A fair amount of agreement was obtained between the experimental and theoretical results. Both indicated that natural sea water is greatly undersaturated with respect to all thirteen metals tested.

This work involved a number of assumptions, particularly as to the composition of the solid phases formed. It has since been pointed out that most of the trace metals do not occur in marine sediments as simple compounds, but exist as substituents in the compounds of more abundant metals with similar properties. This brings about a reduction in the solubility product which may amount to several orders of magnitude (Sillén, 1963). When this is taken into account it appears that the concentrations in sea water of such metals as cobalt and lead could be under solubility control. Sea water is close to saturation with respect to pure barium sulphate (Burton *et al.*, 1968) and it is probable that solubility plays a part in controlling the concentration of barium in sea water.

Krauskopf's investigations of adsorption revealed that hydrous iron and manganese oxides were the strongest adsorbents and that copper, zinc, lead and mercury were the metals most strongly adsorbed. He concluded that adsorption provides an adequate mechanism to explain the apparent undersaturation of sea water with respect to most of the metals tested. The suspended solids of river water include adsorbents such as the clay minerals illite and montmorillonite. Kharkar *et al.* (1968) have shown that a substantial proportion of the trace elements cobalt, selenium and silver adsorbed from distilled water by these minerals is desorbed in sea water. If this process operates in nature the effect would be to increase the supply of trace elements to estuarine waters. The ion-exchange properties of clay minerals in estuarine sediments have been discussed by Dobbins *et al.* (1970).

Field studies of radionuclides discharged to the Columbia River in reactor cooling water have shed light on the exchange of trace metals between solution and suspended solids (Perkins *et al.*, 1966). Most of the metals are discharged in a soluble form, but many, including cobalt, manganese and zinc, become associated with particulate matter during transport to the sea. A notable exception is chromium, probably because it is present in the anionic $+6$ oxidation state. Some of these radionuclides have been detected in sediments off the mouth of the Columbia River (Gross, 1966). Williams and Chan (1966) found that the bulk of the particulate iron discharged by the Fraser River was immediately laid down in the delta.

Krauskopf (1956) was unable to account for the observed concentrations of several trace metals in sea water by adsorption control, and suggested biological uptake as an alternative. Living organisms display a high degree of specificity and high concentration factors in their assimilation of trace elements. For example, different individuals of the ascidian *Molgula manhattensis* concentrate vanadium or niobium, but not both (Carlisle, 1958). Other species of ascidian concentrate vanadium to levels approaching one million times greater than those present in sea water. Vanadium was one of the metals for which Krauskopf postulated biological uptake as a control mechanism. Evidence for the organic removal of two others, cobalt and nickel, is presented by Schutz and Turekian (1965). A significant fraction of the particulate matter in estuaries may be removed by filter-feeding animals. Buchan *et al.* (1967) have suggested that mussel beds are responsible for the lower levels of particulate matter found in the Menai Straits during the summer months.

Non-metals also participate in physico-chemical processes which affect their concentrations in estuarine waters. The problem of silicon removal in estuaries has received considerable attention recently. Bien *et al.* (1958) reported that dissolved silica from the Mississippi River is removed by inorganic processes as it enters the Gulf of Mexico. From laboratory experiments they concluded that silicate was being adsorbed on particulate matter as it settled to the bottom. Appreciable removal has not been found in Southampton Water (Burton *et al.*, 1970), but some removal has been observed in the Conway estuary (Liss and Spencer, 1970). Wollast and De Broeu (1971) have attributed a similar process in the estuary of the River Scheldt to biological uptake by diatoms.

For many years it has been known that estuarine muds are a reservoir of phosphorus, but the importance of phosphate exchange across the sediment/water interface in buffering the concentration of phosphate in the overlying water has only come to be appreciated slowly. The existence of a seasonal cycle in the quantity of phosphate adsorbed on sediments following that of dissolved phosphate was established in Australian estuaries by Rochford (1951). A marked inverse relationship between total phosphorus in the water and sediments of a bay in Florida was reported by Miller (1952). Carritt and Goodgal (1954) demonstrated the reversible adsorption of dissolved phosphate on suspended sediment from Chesapeake Bay in laboratory experiments. They found maximum adsorption under slightly acid conditions, increasing with temperature and decreasing with salinity. Pomeroy *et al.* (1965) found that the concentration of phosphate in equilibrium with cores of estuarine sediment was less than 1 μg-atom P/l. These findings may explain why phosphate is rarely exhausted by phytoplankton in estuarine waters.

The air/water interface is also concerned with processes affecting the concentration of substances in estuarine waters. In addition to the vitally important exchange of the metabolic gases, oxygen and carbon dioxide, across this interface, and the precipitation-evaporation cycle itself, there is the transport of substances dissolved in rain-water to be considered. It has recently been suggested that a significant quantity of phosphorus may be transferred from the land to coastal waters via rainfall (Reimold and Daiber, 1967). The dissolved phosphorus content of rain collected in Delaware rose from less than 5 μg-atoms P/l in the winter to more than 150 μg-atoms P/l during the summer months. The increase is probably derived from agricultural dust. This mechanism may explain the anomalous seasonal cycle of dissolved inorganic phosphate found in some estuaries on the eastern coast of the United States (Jeffries, 1962). Rain-water is also a source of appreciable quantities of ammonium and nitrate (Holden, 1966).

There is a possibility that photochemical reactions participate in the nitrogen cycle of natural waters. Although Hamilton (1964) could find no evidence of ammonium oxidation in sea water exposed to sunlight, it has recently been suggested by Joussot-Dubien and Kadiri (1970) that singlet oxygen produced in sea water by the interaction of visible radiation with naturally-occurring photosensitive organic compounds might bring about this reaction.

So far in this discussion estuarine circulation has been ignored. The mixing induced by tidal currents and river flow controls the introduction of marine materials to the estuary and the removal of freshwater substances to the sea. One consequence of circulation in stratified estuaries is the creation of a trap for suspended matter. A turbidity maximum is found near the boundary between fresh and salt water in estuaries of this type. Particles entering the estuary with river water sink into more saline water which has a net landward flow. Many of these are later returned to the surface layers by turbulence, augmented by particles from the saline water. The cycle may be repeated several times before the particles escape. This has

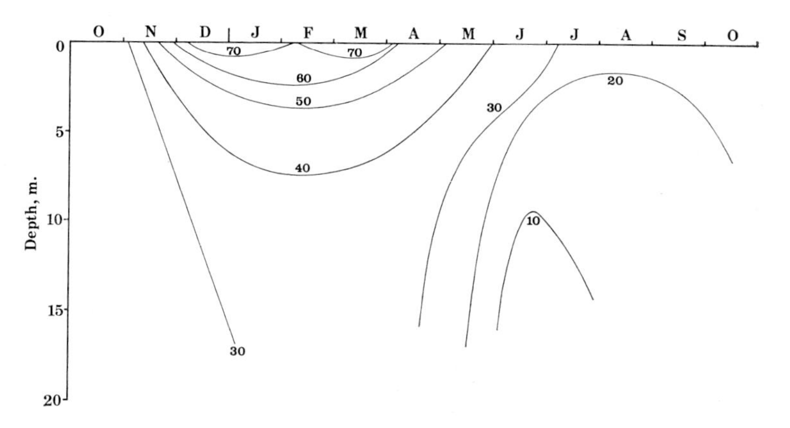

FIG. 8. The annual cycle of nitrate for 1969–70 at Marchwood in Southampton Water. The maximum concentration was 103 μg-atoms N/l.

led to the concept of a 'nutrient trap'. Organic particles are decomposed within the cycle, regenerating nutrients which are used again in the surface layers. In this way nutrients are conserved by the estuary, leading to increased productivity. The relative movement between water and sediments may also play a part in the exchange of material between them. By exposing the sediment to water of varying composition, adsorption equilibria are continually disturbed.

In addition to the tidal cycle, diurnal and seasonal cycles are commonly observed in the chemical composition of estuarine waters. Perhaps the most familiar examples concern parameters involved with photosynthesis, for instance the diurnal cycle in oxygen production and the annual cycle of phosphate and nitrate uptake. River discharge and the input of associated materials are also subject to temporal variations. The combination of these factors leads to a typical pattern of nutrient distribution (Fig. 8).

CONCLUSIONS

Clearly, with so many processes taking place simultaneously, any attempt to disentangle their combined effects, even for a single element, must prove a very difficult task. However, a start can be made by constructing simple mathematical models. Two types of data are necessary for this approach: measurements of the mass of the element present in the various components of the system (water, sediment, organisms, etc.) and estimates of its rate of transfer between the components. A model can be refined as additional information becomes available.

Budgets for nitrogen in the estuaries of the Tyne and the Thames are given in a recent paper by Head (1970). The items on the credit side are the daily supply of nitrogen from river and waste waters, whilst those on the debit side are losses to the sea, the atmosphere (denitrification) and the sediments. In both estuaries most of the nitrogen reaches the sea, but the remainder goes mainly to the atmosphere in the Thames and to the sediments in the Tyne. A budget for phosphate in the Ythan estuary has been published by Leach (1971). It revealed an imbalance between the quantity supplied by the river and the net quantity discharged to the sea over a tidal cycle, which suggested that phosphate was being released from the sediments. The process of model-building is a lengthy one and many estuaries are changing rapidly as a result of human activities. The outstanding question is whether sufficient time remains to establish sound baselines for the prediction of future trends.

From a chemical viewpoint, the picture of estuaries which emerges is of a crossroads between river, sea, atmosphere and sediments. It is the interactions between these phases, coupled with the varied biological activity for which they provide a setting, which are responsible for the wide range of chemical conditions to be found in the estuarine environment.

ACKNOWLEDGEMENTS

I wish to thank my friends in the CERL Marine Biological Laboratory and the Department of Oceanography, Southampton University, for their help in the preparation of this paper, which is published by permission of the Central Electricity Generating Board.

REFERENCES

Baas Becking, L. G. M., Kaplan, I. R. and Moore, D. (1960). 'Limits of the natural environment in terms of pH and oxidation-reduction potentials.' *J. Geol.*, **68,** 243–84.

Barber, R. T. and Ryther, J. H. (1969). 'Organic chelators: factors affecting primary production in the Cromwell Current upwelling.' *J. exp. mar. Biol. Ecol.*, **3,** 191–9.

Barsdate, R. J. (1970). 'Transition metal binding by large molecules in high latitude waters.' In *Organic Matter in Natural Waters*. Ed. D. W. Hood. Pp 485–93. University of Alaska Press.

Bien, G. S., Contois, D. E. and Thomas, W. H. (1958). 'The removal of soluble silica from fresh water entering the sea.' *Geochim. cosmochim. Acta*, **14,** 35–54.

Buchan, S., Floodgate, G. D. and Crisp, D. J. (1967). 'Studies on the seasonal variation of the suspended matter in the Menai Straits. I. The inorganic fraction.' *Limnol. Oceanogr.*, **12,** 419–31.

Burton, J. D., Liss, P. S. and Venugopalan, V. K. (1970). 'The behaviour of dissolved silicon during estuarine mixing. I. Investigations in Southampton Water.' *J. Cons. perm. int. Explor. Mer.*, **33,** 134–40.

Burton, J. D., Marshall, N. J. and Phillips, A. J. (1968). 'Solubility of barium sulphate in sea water.' *Nature, Lond.*, **217,** 834–5.

Carlisle, D. B. (1958). 'Niobium in ascidians.' *Nature, Lond.*, **181,** 933.

Carritt, D. E. and Goodgal, S. (1954). 'Sorption reactions and some ecological implications.' *Deep-Sea Res.*, **1,** 224–43.

Davies, A. G. (1970). 'Iron, chelation and the growth of marine phytoplankton. I. Growth kinetics and chlorophyll production in cultures of the euryhaline flagellate *Dunalliela tertiolecta* under iron-limiting conditions.' *J. mar. biol. Ass. UK*, **50,** 65–86.

Dobbins, D. A., Ragland, P. C. and Johnson, J. D. (1970). 'Water-clay interactions in North Carolina's Pamlico estuary.' *Envir. Sci. Technol.*, **4,** 743–8.

Dodd, J. R. (1965). 'Environmental control of strontium and magnesium in *Mytilus.*' *Geochim. cosmochim. Acta*, **29,** 385–98.

Dragunov, S. S., Zhelokhovtseva, H. H. and Strelkova, E. I. (1948). 'A comparative study of humic acids from soil and peat.' *Pochvovedenie*, 409–20.

Durum, W. H. and Haffty, J. (1963). 'Implications of the minor element content of some major streams of the world.' *Geochim. cosmochim. Acta*, **27,** 1–11.

Foster, P. and Morris, A. W. (1971). 'The seasonal variation of dissolved ionic and organically associated copper in the Menai Straits.' *Deep-Sea Res.*, **18,** 231–6.

Goldman, C. R. (1960). 'Molybdenum as a factor limiting primary productivity in Castle Lake, California.' *Science, N.Y.*, **132,** 1016–7.

Gross, M. G. (1966). 'Distribution of radioactive marine sediment derived from the Columbia River.' *J. geophys. Res.*, **71,** 2017–21.

Hamilton, R. D. (1964). 'Photochemical processes in the inorganic nitrogen cycle of the sea.' *Limnol. Oceanogr.*, **9,** 107–11.

Head, P. C. (1970). 'Discharge of nutrients from estuaries.' *Mar. Pollut. Bull.*, **1,** 138–40.

Head, P. C. and Burton, J. D. (1970). 'Molybdenum in some ocean and estuarine waters.' *J. mar. biol. Ass. UK*, **50,** 439–48.

Hedgpeth, J. W. (1967). 'The sense of the meeting.' In *Estuaries*. Ed. G. H. Lauff. Pp. 707–10. American Association for the Advancement of Science.

Holden, A. V. (1966). 'A chemical study of rain and stream waters in the Scottish Highlands.' *Department of Agriculture and Fisheries for Scotland, Freshwater and Salmon Fisheries Research Series*, No. 37.

Jeffries, H. P. (1962). 'The atypical phosphate cycle of estuaries in relation to benthic metabolism. In *The Environmental Chemistry of Marine Sediments*. Ed. N. Marshall. Pp. 58–68. Narragansett Marine Laboratory, University of Rhode Island.

Jerlov, N. G. (1955). 'Factors influencing the transparency of the Baltic waters.' *Meddn oceanogr. Inst. Göteborg*, **25,** 1–19.

Johnston, R. (1964). 'Sea water, the natural medium of phytoplankton. II. Trace metals and chelation, and general discussion.' *J. mar. biol. Ass. UK*, **44,** 87–109.

Joussot-Dubien, J. and Kadiri, A. (1970). 'Photosensitised oxidation of ammonia by singlet oxygen in aqueous solution and in seawater.' *Nature, Lond.*, **227,** 700–1.

Ketchum, B. H. (1969). 'Eutrophication of estuaries.' In *Eutrophication: causes, consequences, correctives*. Pp. 197–209. National Academy of Sciences, Washington, D.C.

Kharkar, D. P., Turekian, K. K. and Bertine, K. K. (1968). 'Stream supply of dissolved silver, molybdenum, antimony, selenium, chromium, cobalt, rubidium and cesium to the oceans.' *Geochim. cosmochim. Acta*, **32,** 285–98.

Krauskopf, K. B. (1956). 'Factors controlling the concentrations of thirteen rare metals in sea-water.' *Geochim. cosmochim. Acta*, **9,** 1–32B.

Leach, J. H. (1971). 'Hydrology of the Ythan estuary with reference to distribution of major nutrients and detritus.' *J. mar. biol. Ass. UK*, **51,** 137–57.

Liss, P. S. and Spencer, C. P. (1970). 'Abiological processes in the removal of silicate from sea water.' *Geochim. cosmochim. Acta*, **34,** 1073–88.

Livingstone, D. A. (1963). 'Chemical composition of rivers and lakes.' *US Geological Survey Professional Paper*, 440-G.

Lloyd, R. and Herbert, D. W. M. (1962). 'The effect of the environment on the toxicity of poisons to fish.' *Instn. publ. Hlth. Engrs. J.*, **61,** 132–45.

Miller, S. M. (1952). 'Phosphorus exchange in a sub-tropical marine basin.' *Bull. mar. Sci. Gulf Caribb.*, **1,** 257–65.

Nelson, T. C. (1947). 'Some contributions from the land in determining conditions of life in the sea.' *Ecol. Monogr.*, **17,** 337–46.

Odier, M. and Plichon, V. (1971). 'Le cuivre en solution dans l'eau de mer: forme chimique et dosage.' *Analytica chim. Acta*, **55,** 209–20.

Perkins, R. W., Nelson, J. L. and Haushild, W. L. (1966). 'Behavior and transport of radionuclides in the Columbia River between Hanford and Vancouver, Washington. *Limnol. Oceanogr.*, **11,** 235–48.

Pomeroy, L. R., Smith, E. E. and Grant, C. M. (1965). 'The exchange of phosphate between estuarine water and sediments.' *Limnol. Oceanogr.*, **10,** 167–72.

Prakash, A. and Rashid, M. A. (1968). 'Influence of humic substances on the growth of marine phytoplankton: dinoflagellates.' *Limnol. Oceanogr.*, **13,** 598–606.

Reimold, R. J. and Daiber, F. C. (1967). 'Eutrophication of estuarine areas by rainwater.' *Chesapeake Sci.*, **8,** 132–3.

Riley, J. P. and Chester, R. (1971). *Introduction to Marine Chemistry.* London and New York: Academic Press.

Rochford, D. J. (1951). 'Studies in Australian estuarine hydrology. I. Introductory and comparative features.' *Aust. J. mar. Freshwat. Res.*, **2,** 1–116.

Rucker, J. B. and Valentine, J. W. (1961). 'Salinity response of trace element concentration in *Crassostrea virginica*.' *Nature, Lond.*, **190,** 1099–1100.

Ryther, J. H. and Dunstan, W. M. (1971). 'Nitrogen, phosphorus, and eutrophication in the coastal marine environment.' *Science, N.Y.*, **171,** 1008–13.

Ryther, J. H. and Kramer, D. D. (1961). 'Relative iron requirement of some coastal and offshore plankton algae.' *Ecology*, **42,** 444–6.

Sakamoto, W. (1968). 'Study on the turbidity in estuary (II) observations of coagulation and settling processes of particles in the boundary of fresh and saline water.' *Bull. Fac. Fish. Hokkaido Univ.*, **18,** 317–27.

Schutz, D. F. and Turekian, K. K. (1965). 'The investigation of the geographical and vertical distribution of several trace elements in sea water using neutron activation analysis.' *Geochim. cosmochim. Acta*, **29,** 259–313.

Shapiro, J. (1967). 'Yellow organic acids of lake water: Differences in their composition and behaviour.' In *Chemical Environment in the Aquatic*

Habitat. Ed. H. L. Golterman and R. S. Clymo. Pp. 202–16. Koninklijke Nederlandse Akademie van Wetenschappen.

Sheldon, R. W., Evelyn, T. P. T. and Parsons, T. R. (1967). 'On the occurrence and formation of small particles in seawater.' *Limnol. Oceanogr.*, **12,** 367–75.

Sheldon, R. W. and Sutcliffe, W. H. (1969). 'Retention of marine particles by screens and filters.' *Limnol. Oceanogr.*, **14,** 441–4.

Sieburth, J. McN. and Jensen, A. (1968). 'Studies on algal substances in the sea. I. Gelbstoff (humic material) in terrestrial and marine waters.' *J. exp. mar. Biol. Ecol.*, **2,** 174–89.

Sillén, L. G. (1963). 'How has sea water got its present composition?' *Svensk kem. Tidskr.*, **75,** 161–77.

Spencer, C. P. (1957). 'Utilisation of trace elements by marine unicellular algae.' *J. gen. Microbiol.*, **16,** 282–5.

Steemann Nielsen, E. and Wium-Andersen, S. (1970). 'Copper ions as poison in the sea and in freshwater.' *Mar. Biol.*, **6,** 93–7.

Stiff, M. J. (1971a). 'Copper/bicarbonate equilibria in solutions of bicarbonate ion at concentrations similar to those found in natural water.' *Wat. Res.*, **5,** 171–6.

Stiff, M. J. (1971b). 'The chemical states of copper in polluted fresh water and a scheme of analysis to differentiate them.' *Wat. Res.*, **5,** 585–99.

Vishniac, H. S. and Riley, G. A. (1961). 'Cobalamin and thiamine in Long Island Sound: patterns of distribution and ecological significance.' *Limnol. Oceanogr.*, **6,** 36–41.

Williams, P. M. and Chan, K. S. (1966). 'Distribution and speciation of iron in natural waters: transition from river water to a marine environment, British Columbia, Canada.' *J. Fish Res. Bd. Can.*, **23,** 575–93.

Wollast, R. and De Broeu, F. (1971). 'Study of the behaviour of dissolved silica in the estuary of the Scheldt.' *Geochim. cosmochim. Acta*, **35,** 613–20.

4

Physiological Problems for Animal Life in Estuaries

L. C. Beadle

As an environment for aquatic life estuaries are certainly unique. Not only are they zones of transition between two very different environments, sea and fresh water, but, owing to tides and to fluctuations in the downflow of fresh water, conditions at a particular point may oscillate widely and very frequently. Some small bodies of water in arid regions, that oscillate even more widely between fresh, highly saline water and complete desiccation, usually change sufficiently slowly to allow a series of organisms each to complete its life-history under more or less stable conditions. This is not so in estuaries where each organism is subjected to the twice daily tidal oscillations. It is hardly surprising that relatively few species have managed to adapt to these rapidly fluctuating conditions (Fig. 1). If, however, we are right in supposing that most fresh water organisms, other than insects and pulmonate molluscs, are derived from marine ancestors, the brackish waters of estuaries must, through the ages, have supported a large number of species.*

Ecologically the most influential feature is the rapid oscillation in salinity and chemical composition of the water. Nevertheless, the associated changes in flow-rate, depth and temperature are important both directly and in their effects on the structure and composition of the sediments. The amplitude of the oscillations at a given point and the length of river subject to estuarine conditions will of course depend on the gradient and on the volume and rate of downflow from the river.

Some of the land bordering many estuaries is sufficiently low-lying to be flooded with brackish water by the highest tides. The salt-marshes so formed may become more saline than the sea and

* For reviews of ecological work on estuaries and brackish waters, see Emery and Stevenson (1957), Hedgpeth (1957), Segerstrale (1957), Remane (in Remane and Schlieper, 1958), Kinne (1964a) and a general biological account by Green (1968).

conditions become those of inland hypersaline waters with their characteristic fauna and flora.

Another biologically important and characteristic feature of many estuaries, much enhanced by man-made organic pollution, is the particulate organic matter brought down from the river catchments

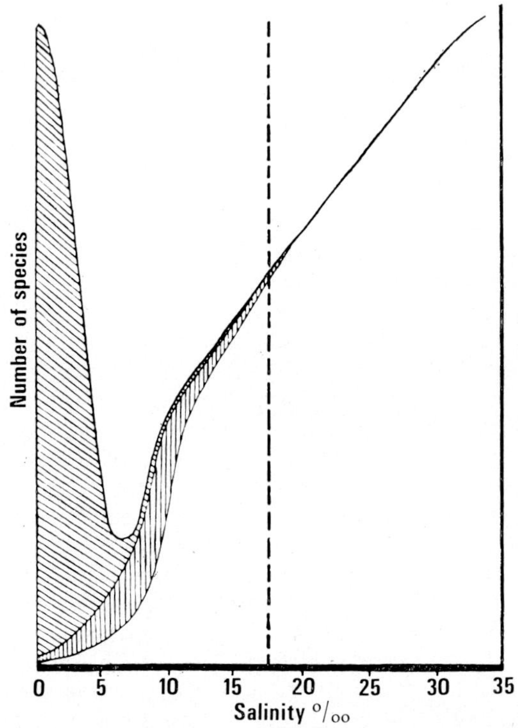

FIG. 1. Remane's generalised representation of the proportions of fresh water (diagonal lines), brackish water (vertical lines) and marine (unshaded) species, based on the Baltic (after Remane and Schlieper, 1958). This pattern, with different relative co-ordinates, is characteristic of all estuaries, *see* for example the Rivers Tees and Tay (Alexander, Southgate and Bassindale, 1935).

and deposited in the sediments. Here it comes into contact with a high concentration of salts, especially sulphate, to provide conditions for vigorous bacterial reduction, so that below the surface sediments are commonly anaerobic and charged with hydrogen sulphide.

Adaptation to any of these estuarine features demands physiological modifications. I propose, however, to confine attention to the physiology of adaptation to the ecologically dominant fluctuations in salinity. This can be approached at different levels—from the whole organism to the intracellular components of the regulatory mechanisms, but I intend to pick out some of what appear to me to be the aspects most directly relevant to ecology.

Though the 'milieu interieur' of many organisms can, as we now know, fluctuate more widely than was suspected by the French physiologist Claude Bernard, who, in the 1870s, first developed the concept of what is now called homeostasis, regulatory mechanisms working between certain limits are one of the necessities for existence in all environments. We are here concerned with the maintenance of a physiologically suitable internal environment with respect to water and salts in the face of wide and rapid external changes.* The problem cannot be solved by a passively impermeable covering because all aquatic organisms (except for dormant and resistant stages) must of necessity 'leak'. Their metabolism involves a continual passage in and out of respiratory gases, water, inorganic ions and small organic molecules. Some or all of the surface of the body and of those tubes and cavities that are in direct communication with the exterior (gut and excretory organs) must remain permeable to these materials. Control of the composition of the interior is therefore a dynamic process working continuously against the normal course of diffusion.

The relations between the tissues of aquatic organisms and the external water are of two kinds. In the case of the protozoa, small algae and many coelenterate animals the cells are for the most part in direct or very close contact with the water, and the control mechanisms are concerned with the regulation of ion and water exchanges directly between the cell interior and the external medium. On the other hand, in the larger multicellular animals such as annelid worms, molluscs, crustacea, echinoderms and chordates, most of the living tissues are bathed by a body-fluid which acts as an intermediary and 'buffers' the effects of environmental fluctuations. The control of exchange between body-fluid and water is performed by relatively few special cells situated on the body surface and in the gut and excretory organs.

Since the body-fluids of the coelomate animals are easy to extract in analysable quantities it is not surprising that the existence of osmoregulatory mechanisms was first demonstrated in brackish water crustacea early in this century. In 1904 Frédericq, by measurements of freezing point, demonstrated hypertonic regulation of the blood of the crab *Carcinus maenas* when placed in dilute sea water.

* For reviews of work on osmoregulation in brackish water animals and on its relevance to ecology see Beadle (1943; 1957), Lockwood (1962), Potts and Parry (1964), Robertson (1957), Schlieper (in Remane and Schlieper, 1958) and Shaw (1960).

From then on the comparative osmotic pressures of the external water and body-fluids of many marine and brackish water coelomate animals were determined by measurement of freezing point or vapour pressure. Some of these are set out in Fig. 2. The most striking feature to emerge from these curves is the very great differences between the level at which different species manage to hold the

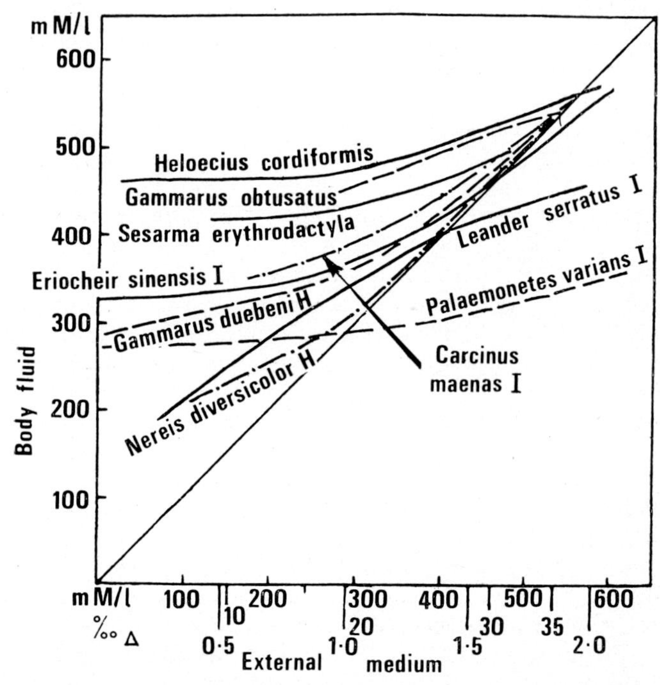

FIG. 2. Relation of the external to the internal medium in various brackish water animals. The left-hand ends of the curves show the approximate low salinity tolerance limits. 'H' and 'I' indicate species that are known to produce hypotonic and isotonic urine respectively in dilute seawater. (Modified from Beadle, 1943.)

osmotic pressure of their body-fluids when the outside water is diluted. It follows that the tissues of some species (*e.g. Nereis diversicolor*) are subject to much greater changes in the osmotic pressure of the body-fluid than others (*e.g. Carcinus maenas*) under the fluctuating conditions characteristic of estuaries. This raises the problem of the adaptation of the tissues to changes in the composition of the body-fluid. There are in fact a number of marine animals, such as the lugworm *Arenicola marina* and the mussel *Mytilus edulis*, that have no powers of osmoregulation but nevertheless penetrate into regions of estuaries where the salinity is periodically reduced to twenty to thirty per cent that of sea water.

Study of the nature of the regulating mechanisms has depended mainly upon the development during the past forty years of ultra-microanalytical methods. It had previously been made clear from analyses of, for example, glandular secretions, urine and the surrounding water that living cells habitually move substances, particularly ions, against gradients of concentration. Moreover 'active transport' of ions is a property of all cells and is not confined to those specialising in absorbing or secreting particular substances. This is true even of the tissue cells of higher animals that are in osmotic equilibrium with the body-fluid surrounding them. The maintenance of ionic disequilibria across the cell membrane is essential to the functioning of all cells. It has recently been shown that, in brackish water animals whose body-fluid concentration alters in response to external salinity changes, there is a re-adjustment of the ionic relations across the cell membranes and the cell contents are brought into osmotic equilibrium with the surrounding fluid partly by adjustment of water and inorganic ions, and partly by alteration in the concentration of free amino acids (Shaw, 1958; Potts, 1958; Florkin and Schoffeniels, 1969).

The component of osmoregulatory mechanisms in aquatic animals first to be discovered was the production of hypotonic urine by the frog in fresh water. This is done by active reabsorption of ions from the filtrate in the kidney, and compensates in part for the loss of ions by diffusion into the water. It seems to be universal in fresh water animals possessing excretory organs but is absent from most, but not all, brackish water species whose urine is isotonic with the blood even when the latter is kept hypertonic to the external diluted sea water. Though uptake of ions from the food certainly makes a contribution, the fact that so many aquatic animals can be kept in a healthy condition for very long periods without food suggests another important source of ions.

It was first shown in the 1930s by Nagel that *Carcinus maenas* in diluted sea water takes up chloride through the gills against a concentration gradient. Krogh then demonstrated that active absorption of ions through part at least of the body surface is a universal feature of brackish and fresh water animals. Since 1945, with developments in techniques of microanalysis, radioisotopes and electron microscopy, much light has been thrown on the details of the mechanisms. There is little doubt that active uptake of ions from the water is the principal means which enabled marine animals to invade brackish and fresh water. Hypotonic urine production was an extra refinement that has facilitated life in fresh water. In addition, the loss of ions has been reduced in brackish and fresh water animals, by a decreased permeability of much of the body surface, thus taking some of the strain from the regulatory mechanism.

Even marine invertebrates whose body-fluids are isotonic with sea water in fact maintain ionic disequilibria across the body wall by active transport, and there is a constant flux of ions in both directions. This is true also of marine protozoa and coelenterates in which the exchange is direct between sea water and cell interior. Active transport was therefore not a new faculty associated with invasion of brackish waters. Marine animals are, to this extent, pre-adapted, and it can be imagined that relatively small changes in the intensity of these processes could maintain an internal medium hypertonic to the external water.

Reversal of the net direction of active transport of ions has enabled some animals to live in hypersaline water in which they maintain hypotonic body-fluids. The teleost fish, which we have reason to suppose originated in fresh water, have re-invaded the sea in which they keep the osmotic pressure of their blood well below that of the seawater. The anadromous species, such as eels and salmonids, adjust the direction of their regulatory mechanisms as they move up or down estuaries. A few euryhaline invertebrates (*e.g.* the prawn *Palaemonetes varians*, Fig. 2) also possess a two-way osmoregulatory mechanism.

The basic facts outlined above concerning the nature of the osmoregulatory mechanisms have been discovered from experiments on animals brought to equilibrium with water of different salinities. What demands on these mechanisms are made by the very high rate and frequency of salinity change, which, as already stated, is a major distinguishing feature of estuarine conditions? It is well known that, up to a point, the range of salinity, as of other conditions such as temperature, which an animal will tolerate can be increased by gradual acclimatisation (Kinne, 1964b). It would be interesting to discover just why such an adaptive change in the regulatory mechanism, if possible at all, can only be accomplished when the environment alters at less than a certain rate. And what in detail is added to the mechanism when a genetic change gives a permanent shift to the limits of the range of tolerance?

It is in fact the *rate* of non-genetic adaptation which is at a premium for organisms in the rapidly fluctuating conditions in estuaries. Relevant in this connection are the experiments of Wells and Ledingham (1940) and of Bogucki and Wojtczak (1962) on isolated rhythmic muscle preparations of *Nereis diversicolor* and *Arenicola marina*, and by Pilgrim (1953) on the isolated heart of *Mytilus edulis*. *Nereis* has some powers of osmoregulation (Fig. 2) but the other two have none, but nevertheless penetrate brackish waters. A sudden fall in the salinity of the medium causes a paralytic shock in the isolated muscle, which, after a time, recovers normal contraction. If, however, the medium is diluted sufficiently slowly to the same final level, the muscle remains active throughout. In the case of *Arenicola* it was shown that the dilution curve of the body-fluids of an intact animal, transferred suddenly to water of the same low salinity, follows a time course that protects the muscles from shock. The body wall acts as a 'buffer' retarding the rate of internal dilution. Besides discovering the intracellular events associated with shock and recovery, it would be interesting to extend this type of experiment both to other purely marine animals and to brackish water species whose body-fluids are regulated to a higher and more constant level of osmotic pressure. It might emerge that the wide range of salinity tolerance of the tissues of animals, such as *Arenicola*, is restricted to non-regulating brackish water species. On the other hand, this may be a necessary step in the invasion of estuaries, which is later superseded by an osmoregulatory mechanism without which life in very low salinities and in fresh water is impossible.

There are several instances of species with populations living in different salinity ranges between sea and fresh water. Such species as have been investigated show, as might be expected, powers of ion-uptake and of reducing the salt-loss from the excretory organs to be inversely related to the salinity of the water in which they live. Some, such as the fresh and brackish water forms of the Baltic isopod crustacean *Mesidotea entomon*, appear to be genetically distinct physiological races since the brackish form is unable to live in fresh water (Lockwood and Croghan, 1957). On the other hand, Sutcliffe (1971) suggests that the various brackish and fresh-water populations of the amphipod *Gammarus duebeni* in Britain and Ireland are not genetically distinct because the differences in their powers of sodium uptake can be induced by prolonged acclimatisation. Whatever the true relative genetic status of these populations, they and the many very closely related species in brackish and fresh water (*e.g. Nereis diversicolor* and *N. limnicola*) are surely evidence of the continuing invasion of fresh waters (via estuaries) from the sea, in which both non-genetic adaptation and genetic selection are playing their parts. There are many estuaries in the world, especially in the tropics, where new and interesting studies could be made on osmotic and ionic regulation in relation to ecology and to the evolution of fresh water animals such as those of Barnes (1967) in Queensland, Australia. For example, there are the clupeid fishes of the West African estuaries and rivers that have penetrated far inland, even to

Lake Tanganyika 1 500 km from the coast. Many other such examples are well known.

A knowledge of the nature of the mechanisms by which animals can cope with the overriding ecological feature of estuaries (salinity fluctuations) is obviously needed for an understanding of their relations with the natural environment; but it is clearly not the only necessary consideration. Temperature is a particularly important variable which can affect estuarine animals both directly, as on development and growth, and indirectly through the osmoregulatory mechanisms. Indeed, salinity changes too can have effects on growth and development. The separate and combined effects of temperature and salinity on marine and brackish water animals have been studied and reviewed by Kinne (1963; 1964b).

The oxygen supply in an unpolluted estuary is usually adequate except below the surface of organic sediments where the inhabitants may require behavioural or metabolic adaptation to a low oxygen level and to reducing conditions. The general metabolic effect of oxygen deficiency is manifested in many ways including in a deterioration of the osmoregulatory mechanisms. Organic pollution has now brought oxygen deficiency to a status of major importance.

It is obvious, though sometimes forgotten, that establishment and maintenance of a species in a given environment depends upon the successful adaptation of all stages of the life history, unless a particular stage can migrate temporarily into another environment to which it is better adapted. From the small amount of work that has been done in this field with estuarine animals it appears that the salinity tolerance range of eggs and early stages is more restricted than that of the adults. This is true of *Nereis diversicolor* whose adults, by virtue of their osmoregulation, can tolerate nearly fresh water in which the eggs and larvae rapidly succumb. It seems that the Californian *Nereis limnicola* is adapted to fresh water partly through internal self-fertilisation and viviparity (Smith, 1964). Another interesting case is *Egeria radiata*, the edible fresh-water 'clam' of the lower Volta River in Ghana. It belongs to the predominantly marine order Tellinaceae. It is restricted to the lower river because its eggs and larvae can survive and develop only in slightly brackish water which penetrates farther up when the river is low during the dry season (Purchon, 1963). For most fully fresh-water animals, however, osmotic independence has been achieved by the eggs and larvae.

Since the techniques for investigating the concentrations and movements of ions are very sensitive, they might be put to practical use in the study of the effects of pollution. It is likely that incipient ill health of an organism would be reflected in small but detectable abnormalities in the ion fluxes before any other symptoms had appeared. It would even be possible to recognise a case in which the permeability of the body surface had been lowered by a pollutant and the consequent increased rate of loss of, say, sodium was being compensated by a more rapid uptake. Such an animal, though still healthy, is nevertheless rather closer to the limits of its endurance. One stage in the life history might well be found to be in greater danger than others.

REFERENCES

Alexander, W. B., Southgate, B. A. and Bassindale, R. (1935). 'Survey of the River Tees. Part II—The estuary, chemical and biological. *Dept. Sci. Indust. Res., Water Pollut. Res. Tech. Paper*, No. 5, p. 171.

Barnes, R. S. K. (1967). 'The osmotic behaviour of a number of grapsoid crabs with respect to their differential penetration of an estuarine system.' *J. Exp. Biol.*, **47,** 535–51.

Beadle, L. C. (1943). 'Osmotic regulation and the faunas of inland waters.' *Biol. Rev.*, **18,** 172–83.

Beadle, L. C. (1957). 'Osmotic and ionic regulation in aquatic animals.' *Ann. Rev. Physiol.*, **19,** 329–58.

Bogucki, M. and Wojtczak, A. (1962). 'Contractility of isolated muscles of *Nereis diversicolor* cultured in hypotonic media.' *Polsk. Arch. Hydrobiol.*, **10,** 231–9.

Emery, K. O. and Stevenson, R. E. (1957). 'Estuaries and Lagoons. I. Physical and chemical characteristics.' *Geol. Soc. Am. Mem.*, **67,** Vol. 1, 673–93.

Florkin, M. and Schoffeniels, E. (1969). *Molecular Approaches to Ecology.* Chapter 6. London and New York: Academic Press.

Green, J. (1968). *The Biology of Estuarine Animals.* London: Sidgwick & Jackson.

Hedgpeth, J. W. (1957). 'Estuaries and Lagoons. II. Biological aspects.' *Geol. Soc. Am. Mem.*, **67,** Vol. 1, 693–750.

Kinne, O. (1963). 'Non-genetic adaptation to temperature and salinity.' *Helgol. Wiss. Meeresunters*, **9,** 433–58.

Kinne, O. (1964a). 'Physiologische und Ökologische Aspekte des lebens in Aestuarien.' *Helgol. Wiss. Meeresunters*, **11,** 131–56.

Kinne, O. (1964b). 'The effects of temperature and salinity on marine and brackish water animals. II. Salinity and temperature combinations.' *Oceanogr. Mar. Biol. Ann. Rev.*, **2,** 281–339.

Lockwood, A. P. M. (1962). 'The osmoregulation of Crustacea.' *Biol. Rev.*, **37,** 287–305.

Lockwood, A. P. M. and Croghan, P. C. (1957). 'The chloride regulation of the brackish and freshwater races of *Mesidotea entomon* (L).' *J. Exp. Biol.*, **34,** 253–8.

Pilgrim, R. L. C. (1953). 'Osmotic relations in molluscan contractile tissues. I. Isolated ventricle strip preparations from Lamellibranchs

(*Mytilus edulis* L., *Ostrea edulis* L., *Anodonta cygnea* L.). *J. Exp. Biol.*, **30,** 318–30.

Potts, W. T. W. (1958). 'The inorganic and aminoacid composition of some lamellibranch muscles.' *J. Exp. Biol.*, **35,** 749–64.

Potts, W. T. W. and Parry, G. (1964). *Osmotic and Ionic Regulation in Animals*. Oxford: Pergamon Press.

Purchon, R. D. (1963). 'A note on the biology of *Egeria radiata* (Bivalvia, Donacidae).' *Proc. Malac. Soc. Lond.*, **35,** 251–71.

Remane, A. and Schlieper, C. (1958). *Die Biologie des Brackwassers. Die Binnengewasser*, **22,** p. 348.

Robertson, J. D. (1957). 'Osmotic and ionic regulation in aquatic invertebrates.' In *Recent Advances in Invertebrate Physiology*. University of Oregon, 229–46.

Segerstråle, S. G. (1957). 'Baltic Sea'. *Geol. Soc. Am. Mem.*, **67,** Vol. 1, 751–800.

Shaw, J. (1958). 'Osmoregulation in the muscle fibres of *Carcinus*.' *J. Exp. Biol.*, **35,** 920–9.

Shaw, J. (1960). 'The mechanics of osmoregulation.' In *Comparative Biochemistry*, Eds. E. M. Florkin and H. S. Mason, **2,** 479–518.

Smith, R. I. (1964). 'On the early development of *Nereis diversicolor* in different salinities. *J. Morph.*, **114,** 437–64.

Sutcliffe, D. W. (1971). 'Sodium influx and loss in freshwater and brackish water populations of the amphipod, *Gammarus duebeni* Lilljeborg.' *J. Exp. Biol.*, **54,** 255–68.

Wells, G. P. and Ledingham, J. C. (1940). 'Physiological effects of a hypotonic environment. I. The action of hypotonic salines on isolated rhythmic preparations from polychaete worms (*Arenicola marina, Nereis diversicolor, Perinereis cultrifera*). *J. Exp. Biol.*, **17,** 337–52.

5

Aspects of Salt-Marsh Ecology with Particular Reference to Inorganic Plant Nutrition

R. L. Jefferies

INTRODUCTION

Estuaries have been defined as semi-enclosed coastal bodies of water which have a free connection with the open sea and within which sea water is measurably diluted with fresh water derived from land drainage (Pritchard, 1967). They are therefore characterised by relatively unstable conditions which reflect the transient environmental gradients that occur in this type of ecosystem. Salt-marshes are formed mainly in the upper inter-tidal areas where there is an appreciable body of mud and sand and where currents usually are slack and waves slight. They have been defined by various authors as communities which occur between the upper limit of high water of spring tides and the upper limit of high water of neap tides (Chapman, 1960; Pigott, 1969; McLusky, 1971). The fact that marshes normally lie in this tidal range means that they are not subjected to the extreme diurnal fluctuations of salinity associated with the movements of river and estuarine water. In spite of their relative stability the origin and development of salt-marshes is nevertheless strongly dependent on changes taking place elsewhere and salt-marshes in turn play an important role in the cycling of materials in estuaries and coastal sites. As pointed out by Pomeroy (1970) salt-marshes are naturally eutrophic systems where normally there is a large excess of all essential elements and a biologically-driven flux of them through the systems. The net primary productivity of these ecosystems is often high and detritus feeders play an important role in the cycling of materials. The vegetation is composed of halophytes which can withstand immersion in sea water for varying periods and the capacity of these plants to grow and reproduce under these conditions has long been of considerable interest.

The mechanisms whereby halophytes regulate their internal ion content is relevant in an understanding of the nature of salt tolerance in plants, and consequently the first section of this review is concerned with the control of ion movement and the requirement by plants of certain ions for growth. In the second section the ecology of different salt-marsh communities is examined and linked to the considerable biological variation that exists within salt-marshes. The productivity of these communities is discussed in the final section and related to the flow of energy in salt-marsh ecosystems. Inevitably in a review of this type there is some unevenness in the depth of treatment given to topics but in certain cases this largely stems from a lack of available information.

SOME ASPECTS OF THE PHYSIOLOGY AND BIOCHEMISTRY OF HALOPHYTIC PLANTS

Ionic relationships of halophytes

The responses of plants in general to the effects of soil salinity have been discussed in a number of papers or books (Hayward and Bernstein, 1958; Bernstein, 1962; Strogonov, 1964; Slatyer, 1967; Jennings, 1968). Although some of these accounts concentrate on the effects of salinity stress on crop plants rather than halophytic plants, the responses of both groups to the imposition of elevated ionic levels in the rooting medium are often similar although the concentrations at which the responses are elicited may be different. The progressive reductions in growth rate caused by increasing salinity appear to be caused primarily by the effect of excess ion accumulation in the affected plants. Direct osmotic effects, acting through reduced water availability to the plants, appear to be of secondary importance, except as initial responses to the imposition of saline substrates. The elevated levels of ions in tissues result in adverse effects associated with ionic imbalance in the tissues of the plant. A high degree of ion-selectivity and well-regulated ion-transport systems appear to be essential if the plant is to exhibit any degree of tolerance to salinity stress during growth and development. As Gutknecht and Dainty (1969) point out in an excellent and detailed review of the ionic relations of marine algae, the problem of osmotic swelling and lysis has largely been solved in plant cells by the development of membrane systems which are relatively permeable to ions but which possess an outward sodium pump which renders the membrane functionally impermeable to sodium. Plant cells have also evolved a rigid cell wall which mechanically resists the tendency of the cell to swell. A number of general conclusions emerge from the results summarised by Gutknecht and Dainty (1969) which, although based on data from algae, are of relevance in a discussion of the ability of halophytic plants as a group to survive saline conditions. The conclusions are that all cells apparently extrude sodium and most cells actively absorb potassium and chloride ions. The cytoplasm of all cells and the vacuole of most cells is electronegative with respect to the external sea water and both the cytoplasm and sap are generally high in potassium and low in sodium, whereas the sap is high in chloride ions.

Although studies on salt-marsh plants are not so thorough as the investigations described by Gutknecht and Dainty in 1964, preliminary data from our laboratory indicate that their conclusions are also applicable to the cortical cells of roots of *Plantago maritima* and *Triglochin maritima*. The active extrusion of sodium, however, has been examined in a number of salt-marsh plants. Arisz *et al.* (1955) and Hill (1967a, b; 1970a, b) have made detailed studies of the salt glands in *Limonium*. The extrusion of sodium from these glands is subject to metabolic control and operates in response to changes in the concentration of sodium in the external medium. In a recent review Lüttge (1971) has discussed the structure and function of plant glands including the glands of *Atriplex* and *Limonium*. Extremely high effluxes of ions have been reported by Scholander *et al.* (1962, 1966) and by Atkinson *et al.* (1967) in the different species of mangrove. The latter authors report that the observed maximum efflux of chloride from the leaves of *Aegiatites* was approximately 90 picomoles/cm^2/s, which corresponds to a flux of about 5 nanomoles/cm^2/s over the total cross-section of the glands, a very high ionic flux for plants. The results of studies on the efflux of ions from algal cells and the cells of higher plants suggest that halophytic plants are extremely efficient at exporting sodium and other ions from the cells against the electro-chemical gradient and that the rate of movement of ions from the tissue appears to be linked with the degree of salinity stress which the plants experience. One interesting question for which at present there is little information is how far intolerance to high sodium levels is linked with the inability of the sodium pump to cope with enhanced internal levels of sodium when the plant is subjected to stress and as a corollary to this, the extent to which these pumps are constitutive or adaptive. These questions are extremely difficult to solve, partly on account of the necessity to examine the rates of ion movements in relation to the different compartments within the cells and tissues of plants (MacRobbie, 1971).

In higher plants the ways in which toxic levels of ions such as sodium may be avoided often involve long distance transport systems. The transfer of ions from the root to the shoot may be restricted or ions may be exported from the shoots, leaves and roots as previously discussed. In some plants increased succulence may alleviate the stress associated with high concentrations of ions as a result of a dilution effect brought about by an increased water content of leaf cells. In both halophytes and non-halophytes sodium and chloride ions are not exported once they have entered the leaves (Greenway and Thomas, 1965; Greenway, Gunn and Thomas, 1966), whereas potassium may be re-translocated (Greenway and Pitman, 1965). Such data indicate the complex strategies which may be adopted by plants in the event of salinity stress.

As mentioned above there is a considerable amount of evidence, particularly from studies on algal cells, indicating that in plants from brackish and marine habitats the cytoplasm of the cells is generally high in potassium and relatively low in sodium. An extreme case is seen in halophytic bacteria. Christian and Waltho (1961) established a relationship between the potassium content of the cells of bacteria and their tolerance towards sodium chloride. The higher the salt tolerance of the bacteria the higher was the internal potassium content. For example, cells of *Halobacterium salinarium* grown in 4 molar sodium chloride and 0·3 molar potassium chloride contained 4·5 molar potassium internally. Although salt-marsh vegetation rarely encounters these extreme conditions, one of the consequences of sodium extrusion from plants is a high internal potassium environment, which in turn has resulted in the adaption of enzyme systems to high potassium levels (Gutknecht and Dainty, 1969). Baxter and Gibbons (1954, 1956, 1957) and Holmes and Halvorson (1965) discovered that the enzymes of extreme halophilic bacteria are adapted to function at the very high salt concentrations found within cells. Recent investigations of the activity of a number of enzymes such as malate dehydrogenase from the roots and shoots of salt-marsh plants, compared with glycophytic plants, indicate that the enzymes from the halophytes show maximum activity at higher ionic concentrations than the corresponding enzymes of plants from non-saline environments (Sims, Stewart and Folkes, 1972; Raju and Jefferies, 1972). These enzyme systems are strongly activated by a range of cations including sodium, potassium, magnesium and calcium and together they appear to play an important role in keeping the enzymes in an activated state. The cations also appear to play an important role in stabilising the enzyme systems (Larsen, 1967). Although many enzyme systems are activated by potassium (Evans and Sorger, 1966), the requirement for potassium rarely appears to be highly specific, as the majority of enzyme systems examined from halophyte tissues are activated by a number of other ions. This raises the question of why salt tolerance appears to be linked with high internal potassium concentrations in the cells of halophytes. Although there is little evidence available from higher plants, Lubin (1964) has investigated this problem in bacteria. He finds that protein synthesis is unusually sensitive to the presence of a high potassium environment and suggests that this ion plays an important and unique role in the formation of the threefold complex of S-RNA, ribosomes and messenger RNA—a role which cannot be achieved by sodium. If Lubin's suggestion is found to be correct then it may well account for the persistence of high levels of intracellular potassium in cells. Bayley and Kushner (1964) have isolated the ribosomes from *Halobacterium cutirubrum* and reported that they are only stable in a 4 molar K^+ solution which is a specific requirement. In a later paper Bayley (1966) indicates that the chief chemical feature which distinguishes the ribosomes of *Halobacterium salinarium* from those of non-halophilic organisms is the high content of acidic protein and he suggests potassium may be needed in high concentrations to neutralise the negative charges on the acidic protein and RNA, thereby enabling these components to aggregate and function satisfactorily. It is clearly of interest to examine whether these ideas are applicable to halophytic plants.

Ionic and osmotic regulation in plants during changing external environmental conditions, as distinct from conditions where a constant ionic environment exists, have received a limited amount of attention. The ability of plant cells to adjust to changes in external ionic concentrations and yet maintain growth provides some measure of the capacity of the organism to alleviate ionic stress. It is well known that certain algae attached to ships survive daily trips from sea water to fresh water. Intertidal algae withstand desiccation by drying and freezing as well as the dilution of their external solution by rain water. Strains of *Enteromorpha* can survive environmental fluctuations of salinity from 0 to 140‰ (Biebl, 1962). If *E. intestinalis* is transferred suddenly from sea water to fresh water a rapid net outward flux of sodium occurs, together with a slow efflux of potassium and chloride ions. Although the cells are in a swollen state in fresh water growth apparently proceeds at about the normal rate. The ability of *Enteromorpha* to adapt to a range of salinities appears to differ amongst the physiological races of the particular species (Tarr, unpublished). *Valonia* can survive in slowly concentrating sea water provided that adequate external potassium is present and under these changing conditions the potassium concentration of the sap increases from 600 mM to 800 mM (Blinks, 1951). Preliminary results on *Plantago maritima* and *Triglochin maritima* indicate that changes take place in the fluxes of sodium and potassium which are similar to those occurring in *Enteromorpha* as the salinity is altered. For example, there is a rapid net efflux of sodium from the roots of these plants when the sodium concentration in the external solutions is dropped from that of sea water to that of fresh water. In *Triglochin maritima*, an increase in the external sodium level results in an increase in the uptake of potassium by the roots of this species (Parham, 1971). This increased uptake is under metabolic control and is only evident when sodium is at sufficiently high concentrations in the external solution. Epstein (1969) has also examined the uptake of potassium in species of wheat grass subjected to elevated levels of sodium in relation to dual carrier mechanisms which may operate in the roots. The prospects for future work on ionic regulation in plants look extremely interesting, as more detailed studies should provide considerable information and insight on how plants adjust to changes in the ionic environment, yet maintain adequate growth rates.

A topic which has received very little attention during the last decade is the water and osmotic relations of salt-marsh plants. Although a number of general accounts of the effects of salinity on internal water deficits and the movement of water though the soil-plant-air continuum have appeared (*e.g.* Slatyer, 1967), there has been very little new information. However, some comment should be made concerning the matrix potential in soils. Some accounts of the availability of water in saline soils for plant growth only stress the lowering of the free energy of water as a result of osmotic forces and neglect the role of the matrix potential in lowering the water potential of soils. In coastal saline soils which contain a high proportion of silt and clay the matrix potential may be of considerable importance in lowering the free energy of water. It may account for the poor ability of some plants to grow in clay compared with sand even though the apparent osmotic potentials of the soil solutions are similar.

Although Hayward and Wadleigh (1949) postulated that 'salt accumulator' plants can overcome exposure to low osmotic potentials by the absorption of electrolytes, the postulate presupposes that the cells of the plant are tolerant of high ionic concentrations. However, it is well known that plant cells contain a number of charged substances in addition to large quantities of cations and anions. In this context it is of interest that some plants, when they are subject to osmotic stress, contain large quantities of the amino acid proline, which is one of the most soluble of the amino acids and this substance may alleviate some of the need for high internal levels of ions to help maintain adequate water potential gradients both within a cell and within a plant. At present there is little direct evidence that coastal salt-marsh plants contain large quantities of proline but the problem deserves investigation.

Another topic which has received relatively little attention recently is photosynthetic activity and gaseous exchange in temperate salt-marsh plants, particularly when they are subject to tidal action and high levels of salinity. In a recent paper Woodell and Mooney (1970) have shown that the photosynthetic capacity of *Limonium californicum* is little affected by large changes in the osmotic potential of the root environment. It is of interest to know whether other salt-marsh species respond to salinity changes in a similar manner and also the extent to which the two types of photosynthetic carbon fixation, the C_3 type and the C_4 type, operate as appears to occur in some salt-tolerant species of *Atriplex* (Osmond, 1970).

Growth of halophytes in relation to the availability of nitrogen and phosphorus

An aspect of the nutritional relationships of salt-marsh plants which has received some attention recently is the availability of nitrogen and phosphorus in salt-marshes for plant growth. Pigott (1969) has cultured plants of *Salicornia europaea* and *Suaeda maritima* collected from different sites on the north Norfolk coast and supplied the soil cores with nitrogen and phosphate and phosphate alone. The nitrogen was supplied as ammonium nitrate and the phosphorus as sodium phosphate. It was evident that plants of both species present in cores from the upper marsh exhibited a marked growth response to additions of nitrogen and phosphorus whereas the response to an increase of phosphorus alone was much less marked compared with the controls. These plants, which are annuals, do not germinate until May and it is clear that rapid growth can be maintained only if there is a plentiful and available supply of both these elements. There appeared to be a general similarity in the concentration of soluble nitrogen compounds and phosphate at all levels of the marsh and as Pigott points out the significant difference between the bare mud and the upper parts of the marsh is the extent to which the sediment is exploited by the roots of the perennial species already present. He postulated that when nitrogen and phosphorus are added the supply becomes adequate for all plants. As discussed later, on the north Norfolk coast the perennial plants exhibit a rapid growth of the shoot systems during May and this phase of growth may result in the temporary exhaustion of nitrogen in the rooting zone of the soil thereby affecting the growth of these annuals.

The very recent paper of Stewart, Lee and Orebamjo (1972), who have examined the activity of nitrate reductase in plants of *Suaeda maritima*, is particularly relevant here. In all higher plants so far investigated the enzymes concerned in the reduction of nitrate to ammonia have been shown to be adaptive; for example, the enzyme nitrate reductase is induced in the presence of the nitrate ion (Beevers and Hageman, 1969). Stewart *et al.* found that while seedlings of this species are widely distributed over the whole marsh their subsequent growth is dependent on their site of establishment. Seedlings growing in the upper marsh are nitrogen deficient and contain very low amounts of nitrate reductase, whereas in the case of those present at the seaward end of the marsh nitrogen is readily available and the levels of this particular enzyme in the tissues are high. Additions of nitrate to sites in the upper marsh quickly resulted in the appearance of nitrate reductase in the leaf tissues indicating that nitrate, when it is available, is rapidly absorbed by the plants. One problem which is not fully resolved at present is the extent to which plants at the seaward end of the marsh augment their nitrogen supply from nitrogen carried in airborne spray.

Tyler (1967) studied the effects on Baltic shore meadow vegetation of the addition of nitrogen and phosphorus supplied as the following salts, sodium dihydrogen phosphate, sodium nitrate, and ammonium chloride. Whereas the addition of ammonium chloride brought about an increase of thirty per cent in the productivity of the field layer compared with the control plots, when nitrate was added to other plots there was little or no corresponding increase in productivity. The question of how far plants of different salt-marsh species are able to utilise nitrate or ammonium ions has not been fully resolved and more studies are needed to determine the ability of plants at the various stages of their growth cycles to absorb and utilise these ions. Although additions of phosphate to some sites within salt-marsh soils may result in some increase in productivity of the standing crop (Tyler, 1971) it appears that nitrogen is the critical limiting factor for growth in estuaries (Ryther and Dunstan, 1971). In studies on the productivity of *Spartina* marshes, Pomeroy *et al.* (1969) have estimated that there is sufficient phosphorus in the top metre of sediment to meet the requirement of the standing crop for this element for a period of 500 years. Phosphate is readily exchanged between the clay matrix and the bulk water of estuaries and hence there is a very large reservoir of phosphorus in the muds which is partly recycled as a result of exchange at mud–water interfaces. Phosphorus also moves into the *Spartina* crop and is released during the breakdown of plant detritus. The subject of mineral cycling in different ecosystems has been reviewed by Pomeroy (1970). In studies on the turnover of elements in the above-ground biomass of the vegetation of Baltic meadows Tyler (1971) has estimated that only twenty-six per cent of the total uptake of phosphorus by the standing crop was retained in the litter at the end of a season. The 'turnover' curves of many elements are more or less sharply bell-shaped with a maximum level in the standing crop in summer, followed by a rapid decrease due to leaching and/or transport to the below-ground biomass.

However, the curves for some elements such as iron, nickel, lead and aluminium are different, with maxima in late autumn when the above-ground biomass is essentially litter. Studies of this type are few, and there is a need to obtain additional information on the cycling of elements in the different communities within salt-marshes, and where possible relate this movement to the growth of vegetation.

THE ECOLOGY OF SALT-MARSHES

The halophytic element of our flora was much more widespread in inland regions during the Full-glacial period at the end of the Weichselian glaciation than it is today (Bell, 1969). Although some of the sites may have been part of estuaries or brackish systems it is very difficult to associate all the sites with such conditions and hence a more general explanation of the presence of halophytes in inland regions is required. In regions of permafrost, salinity develops because of the impediment to drainage combined with certain climatic conditions (Monoszon, 1964). In Siberia today in permafrost areas saline communities resembling the Weichselian fossil assemblages occur (Cajander, 1903). Unfortunately there is no definite evidence of permafrost associated directly with the plant beds of the British Full-glacial, but Bell postulates that its widespread occurrence was linked with the development of local saline areas in which a halophytic flora grew.

The flora of British salt-marshes today consists of a distinct but limited group of species which are further restricted in Scotland and Ireland. In spite of a number of plantings *Spartina anglica* has not spread extensively on Scottish salt-marshes, and plants such as *Limonium vulgare, L. humile, L. binervosum* and *Halimione portulacoides* are absent or infrequent in Scotland (Gimingham, 1964). Likewise in Ireland, *Limonium vulgare* is absent (Perring and Walters, 1962; Boorman, 1967). As Gimingham (1964) points out, in spite of the limitation and uniformity of the flora a number of distinctive communities exist, and there is considerable interest in the spatial and dynamic relationships of these. However, an experimental approach to the problem of the distribution of particular species within salt-marshes has been largely neglected. The dynamic seral relationships between communities within a marsh or within a region have often been shown in the form of a flow diagram but the assumption that zonation reflects succession does not appear to be always fully justified. One difficulty with such an assumption is that individual plants in salt-marshes may be long lived so that conditions prevailing at the time of establishment bear little similarity to conditions existing on a marsh two or more decades later. Another difficulty relates to the genetic variation shown by many plant species which grow in salt-marshes (Table 1). As long ago as 1944 Gregor presented evidence that *Plantago maritima* consisted of a series of populations adapted to different soil types in maritime habitats. Since that time the existence of considerable

TABLE 1

SALT-MARSH SPECIES COMPOSED OF POPULATIONS WHICH SHOW EDAPHIC ADAPTATION

Species	*Reference*
Agrostis stolonifera	Aston and Bradshaw (1966)
Armeria maritima	Chapman (1960)
Aster tripolium	Chapman (1960)
	Gray (1971)
Halimione portulacoides	Chapman (1960)
	Sharrock (1967)
Plantago maritima	Gregor (1944)
Puccinellia maritima	Gray (unpublished)
Suaeda maritima	Chapman (1960)

genetic variation between populations of salt-marsh species has been established, hence in the light of this evidence the early concepts of zonation and succession as applied to salt-marsh vegetation may require modification.

The existence of such biological variation over extremely short distances indicates that in spite of any gene flow which may occur as a result of the movement of pollen or seeds, selection forces are sufficiently strong to maintain this variation. Antonovics (1968) has discussed the ability of perennial plant populations to withstand the deleterious effects of pollen flow in a study of evolution in plant populations. Most of the variation which has been recognised in salt-marsh populations relates to morphological characteristics, such as seed weight and leaf size and shape, there being little direct evidence of variation in physiological processes amongst populations or individuals of species growing in these saline conditions. Although genetic variation has been recognised, relatively little is known about how this variation is produced and maintained and about the nature of the selective advantages conferred upon populations and individuals. In some respects the problem is similar to that of the evolution of heavy metal tolerance in plant populations in that disruptive selection appears to be operating over comparatively short distances.

The variation within plant communities composed of different species is widespread on marshes. An interesting aspect of the differences in distribution of the common salt-marsh species over distances of a few metres is that they often mirror corresponding differences in the distribution of salt-marsh plants which occur across the whole marsh and which represent a much larger scale of pattern. Evidence of the existence of this point to point variation in salt-marshes results from an analysis of the complex patterns of vegetation that are often found, particularly where the marshes are old. In recent years considerable attention has been given to elucidating the underlying reasons which may account for these complex patterns, studies which have been based largely on the use of multivariate statistical procedures in analysing field data. Marked differences in the distribution of species over short distances of a few metres or less in relation to microtopography have been shown by Packham and Liddle (1970) and Boorman (1971). The detection of these microtopographical differences has been greatly facilitated by the use of a microtopograph devised by Boorman and Woodell (1966), an instrument which enables measurements of the height of the ground to be taken from a fixed datum level.

Examples of studies which attempt to analyse the distribution of selected species in relation to a number of environmental parameters are evident in the work of Sharrock (1967), Dalby (1970), Brereton (1971) and Gray and Bunce (1972). The last two authors have analysed the relationships between soils and vegetation in salt-marshes in Morecambe Bay using a combination of multi-variate statistical procedures. They relate the distribution of plants along a transect to a continuum of developing soils and stress and although various high to low marsh series are apparent it is often not possible to zone areas of saltings in an exact linear sequence.

The analyses reveal that the differences between vegetational types are not directly related to their absolute position above low water mark. Rather, interactions over a period of time between vegetation and soil result in a pedogenetic development which leads to considerable point to point variation of soils and of vegetational types. One of the implications of these results is that interpretation of the types of pattern shown by the vegetation will very much depend on the past and present edaphic and biotic conditions at any given site. In some situations changes may be rapid. Ranwell (1961; 1968) has reported that under conditions of rapid siltation it may take as little as thirty years for a *Spartina* marsh to be completely replaced at its upper limits by *Puccinellia maritima* where the marsh is grazed, or by *Phragmites communis* where the marsh is ungrazed.

Sharrock (1967) examined the distribution of *Halimione portulacoides* in salt-marsh communities in southern England in relation to a range of environmental parameters. As he points out several of these parameters are interacting so that in spite of the use of multivariate statistical procedures it is not always possible to determine the nature of causal relationships in this type of study. The survey showed that plants of this species can tolerate widely differing substrate water contents and *Halimione* occupied every position in the salt-marsh from the upper regions of the Spartinetum to levels higher than that of any other salt-marsh plant. Furthermore, a range of continuous morphological variation in the variety *latifolia* was found between types confined to mud, sand and pebble. The evidence available suggested that this morphological variation was habitat-induced. These results again indicate that this species, in common with many other salt-marsh species, is widely distributed throughout the marsh, that its distribution is not directly correlated with absolute height above Ordnance Datum and above all that individuals of populations of this species exhibit a considerable amount of phenotypic plasticity in exploiting a range of different habitats within a marsh. Sharrock suggests that this morphological variation is linked with interference effects associated with the influence of interspecific density. This last point will be discussed in more detail later. Brereton's study (1971) on populations of *Salicornia europaea* and *Puccinellia maritima* indicates that *Salicornia* tended to occur in highly saline non-waterlogged areas whereas *Puccinellia* occurred in non-saline waterlogged sites. These results were also confirmed experimentally when the effects of waterlogging and salinity on the growth of plants were examined.

In a similar study Boorman (1971) found that *Armeria maritima* and *Plantago maritima* occur on slight eminences whereas *Puccinellia maritima* grows in depressions at Scolt Head Island. He was able to show a negative association between a number of pairs of species including *Plantago maritima* and *Triglochin maritima, Aster tripolium* and *Triglochin maritima, Plantago maritima* and *Halimione portulacoides*.

The results of these papers provide a considerable amount of information on the distribution of populations of species in salt-marshes. They show that the observed vegetational patterns are often complex and that they represent the effects of interactions between plants and soils so that conditions at any one site are closely related to the biological and physico-chemical changes which have taken place at that site over a period of time. The results also imply that the sharp zonations which are often observed at sites where there are continuous environmental gradients are a reflection of the outcome of interactions between individuals of the different species. Hence there is a need for a greater number of studies on intra- and inter-specific competition of plants under conditions similar to those in salt-marshes in order to determine the origin and development of these zonations. It is the comparative biology of the individual species which is probably of significance in analysing the distribution of species in marshes and the way in which, in a given situation, one particular species in a community may face extinction or increase in abundance. This aspect of the ecology of salt-marsh species is of added interest in relation to the known genetic variation within such species. Much of the variation which has been observed in salt-marsh plants may be much more closely linked to the adverse effects of interference from nearby plants than to the influence of the edaphic factors as such.

Examples of the adverse effects of interference between plants of two species under a given set of conditions are seen in the data of Gray (unpublished) and Jefferies (unpublished). In both cases the experiments were designed in a manner similar to the experimental models of de Wit (1960). In these models two species are sown together in various proportions whilst the overall density of the mixture of plants is maintained constant in the different cultures. The behaviour of a species in a pure stand can be compared with its performance in various proportioned mixtures and the mutual interactions of the two species measured. In most cases the parameter frequently measured is the dry weight of the root or shoot and the data are usually plotted as a proportion of the dry weight produced in the pure stand in each case. Such experiments are relatively easy to perform and can provide a considerable amount of information on the capacity of individuals in mixed populations to exaggerate and exploit interspecific differences. Gray grew populations of *Puccinellia maritima* and *Agrostis stolonifera* in pure and mixed stands at three 'tidal' regimes representing different degrees of inundation. Although plants of both species were capable of growing satisfactorily in cultures where additions of sea water were few, if plants of *Agrostis* were present in the cultures, the growth of plants of *Puccinellia* was severely affected. Where additions of sea water were frequent the reverse situation prevailed and the growth of plants of *Agrostis* was markedly checked in mixed cultures.

The author cultured *Plantago maritima* and *Triglochin maritima* in pure and mixed stands at different levels of salinity. Although plants of both species are capable of growing at all salinities, in mixed cultures interspecific differences are exaggerated. The growth of plants of *Triglochin maritima* was much reduced in all cultures when plants of *Plantago maritima* were present, whereas the growth of *Plantago maritima* was only poor when plants were subject to marked intra-specific competition. The stress of density in mixed populations under a given set of environmental conditions creates a hierarchy amongst populations of different species in their capacity to exploit particular habitats (Harper, 1967). It appears likely therefore that the zonation in salt-marshes is in part a measure of this hierarchy.

Perhaps one of the most striking characteristics of salt-marsh vegetation is the large number of species which are perennial. With the notable exceptions of *Suaeda maritima* and some species of *Salicornia* many of the common salt-marsh plants are perennial. Annual plants, although present throughout salt-marshes, are chiefly found in the unstable 'interface' regions such as at the seaward and landward edge of marshes in situations where their colonising ability allows them to exploit these unstable habitats in contrast to perennial plants.

The majority of individuals of perennial plants in marshes appear to be long-lived. Estimates of the age of individual plants of *Limonium vulgare* at Stiffkey, Norfolk, are in excess of forty years. Gimingham (1964) has drawn attention to the considerable age of plants of *Triglochin maritima* in Scottish marshes and the changes which take place as the plant matures. If this longevity, as seems probable, is found to be characteristic of a number of salt-marsh species, it has important implications. First, conditions at the time of establishment of the individual may bear little similarity to the conditions existing two or three decades later—conditions which have been produced, in part, by the presence of the individual at the site over this time period. Secondly, a low death rate of individuals implies a low recruitment rate into the population, at least at the seedling stage. Detailed observations carried out on the high marshes of the north Norfolk coast indicate that seedling establishment over a two-year period was almost totally absent for all the perennial species. Hence, the strategy of the individuals of various species in maintaining themselves under such conditions is to use much of the available resources for vegetative reproduction rather than sexual reproduction. Some sexual reproduction occurs but much of the seed is exported from the immediate vicinity of the parent population as a result of tidal action. Extensive germination of seed of salt-marsh plants may be observed in tidal litter at high tide mark in the early spring, but the mortality of these seedlings is extremely high in summer.

The channelling of resources towards vegetative reproduction in perennial salt-marsh plants can be shown experimentally. Under conditions of high salinity, individual plants of species such as *Plantago maritima* and *Triglochin maritima* fail to flower so that all the available resources are diverted into vegetative reproduction. In addition, under conditions of high intraspecific density stress, flowering also fails to occur in both species even though the salinity of the cultures may be low. The combination of high salinity and high density of individuals in many salt-marsh sites may result in reduced sexual reproduction of certain species. This suggestion is consistent with the observation that many individual plants in salt-marshes flower infrequently. In the case of *Plantago maritima* it is of interest that where plants grow and reproduce in non-saline conditions, they flower extensively but appear to be short lived. These results on the proportion of the plants' output involved in sexual reproduction under conditions of density and salinity stress suggests that in these plants there is a high degree of plasticity, so that under stress a marked shift towards vegetative reproduction and longevity of individuals occurs. More data are required on whether plants of other salt-marsh species exhibit a similar strategy.

The seasonal pattern of these perennial plants appears to be similar, although the evidence at present is based on comparatively few studies. The growth of perennial plants at Stiffkey in Norfolk is characterised by a phase during early summer when rapid growth of the shoot systems occurs. This is sustained over a period of a few weeks but shortly after this phase the standing crop declines. This decline is associated with the development of hypersaline conditions in the rooting zone and possibly an exhaustion of nitrogen in the soil. In addition, high levels of sodium and low levels of potassium occur in the leaves of most of the common salt-marsh plants. The decline in the standing crop is associated with the export of material to below-ground portions of the plants and at present it is uncertain whether these symptoms of ionic imbalance in the tissues are in part the cause or the result of the onset of senescence of the shoot systems. In late summer the accumulation of litter from the standing crop is largely removed as a result of tidal action.

Ranwell *et al.* (1964) have also reported elevated levels of chloride in *Spartina* marshes in Poole Harbour during summer. Tyler (1971) has reported an export of material from the above-ground portions of sea meadow plants in the Baltic to the below-ground portions in high and late summer and if the vegetation is cut during the middle of the summer the above-ground biomass the following year is severely reduced.

Whatever the causes of senescence of the shoot systems of

salt-marsh plants, it is evident that genotypes within the populations which can complete their growth cycle during a relatively short period of time in early and mid-summer are at a selective advantage. For example, with one or two exceptions, at Stiffkey High Marsh both the vegetative growth phase and the sexual reproductive phase of plants are completed in a relatively short period and furthermore there is very little separation in time of the rapid growth phase of plants of the different species during early summer. Tyler (1971) presents similar evidence for the standing crop of the Baltic sea meadows. Studies of the annual growth cycle of *Spartina anglica* by Goodman (1960) and Taylor and Burrows (1968) indicate that the peak in shoot production appears to occur from early May to the end of July, although direct comparison of the data is not possible owing to seasonal differences and differences in sampling procedures. A comparison of the growth rates of different genotypes of salt-marsh species collected from a range of sites and grown under controlled conditions would be of particular interest in establishing whether the short growth phases observed are habitat-induced.

One consequence of this short growth phase is that if the plants are grazed at this stage almost the entire output of seed may be lost. At Stiffkey, the action of rabbits led to the removal of 95% of the flowering heads of *Triglochin maritima* in marked areas. Although a few plants produced a second crop of flowering heads much later in the season, many of the seeds failed to ripen and the inflorescences were short and stunted compared with those produced earlier in the season. The effects of grazing at this stage therefore severely reduced the output of seed from plants of this species even though some

TABLE 2

THE PRINCIPAL SPECIES IN BRITISH SALT-MARSH COMMUNITIES

Ungrazed marshes	*Grazed marshes*
Juncus maritimus and *Festuca rubra*	*Juncus maritimus* and *Festuca rubra*
Armeria maritima, Limonium vulgare and *Plantago maritima*	*Armeria maritima, Festuca rubra* and *Puccinellia maritima*
Halimione portulacoides	*Puccinellia maritima*
Aster tripolium	*Puccinellia maritima*
Phragmites communis and *Scirpus maritimus*	*Puccinellia maritima* and *Atriplex hastata*
Spartina anglica	*Spartina anglica*
Salicornia europaea agg.	*Salicornia europaea* agg.

After Ranwell (1961), Pigott (1969) and Dalby (1970).

plants attempted to produce additional seed. Although grazing may influence the amount of seed formed, it is well known that in many species seed set varies considerably from year to year. Values of between 15 and 92% seed set have been recorded for *Spartina anglica* for particular years (Goodman, 1960; Taylor and Burrows, 1968). As pointed out by Ranwell (1961) as a result of his controlled sheep grazing experiments and later by Pigott (1969) and Dalby (1970), the composition and structure of salt-marsh vegetation is greatly influenced by grazing, particularly by sheep. The principal species in the corresponding zones are shown in Table 2.

Although grazing by sheep is widespread other animals also feed extensively on salt-marshes. Ranwell and Downing (1959) have drawn attention to fluctuations in numbers of Brent geese in relation to their food supply, which is mainly *Zostera nana* and *Enteromorpha*. In addition to birds feeding on salt-marsh vegetation, there is often an abundant mollusc fauna which feeds on algae and plant detritus. MacDonald (1969) has examined the mollusc faunas of the *Spartina-Salicornia* salt-marshes of the North American Pacific Coast. One or two species are widely distributed and very abundant while the remaining species are represented by small numbers of patchily distributed individuals. This distribution pattern was similar between sites which were widely separated. At certain sites very high densities of molluscs such as *Hydrobia ulvae* may be recorded. McLusky (1971) estimates that densities of 42 000/m^2 of this gastropod occur on the Clyde estuary. Since the population dynamics of these mollusc faunas are not well understood at present, there appears to be considerable need for information with regard to the ecology of this group in relation to their spatial and temporal distributions.

Marked habitat selection is shown by species of Acarina in salt-marshes. Luxton (1964) considers the fauna to be halophilic, but some of the members of this group have been collected in a wide variety of other habitats. They appear to exhibit marked food preferences towards different species of soil fungi and Luxton suggests that the striking non-random distribution of individual species of Acarina in salt-marshes may be linked in part to the presence of species of fungi in certain areas of marsh. In a recent paper Duffey (1970) has drawn attention to habitat selection by spiders on a salt-marsh in Gower. In general the spider fauna is not rich and most species are also found elsewhere, particularly in fresh-water marshes. Two species, *Lycosa purbeckensis* and *Erigone longipalpis*, are, however, more characteristic of salt-marshes than other habitats. Stebbings (1971) also reported that the latter species is found in a salt-marsh. From the distribution patterns of spiders Duffey was able to demonstrate the ability of successful species to establish themselves in normally sub-optimal environments, when conditions ameliorated for short periods.

The available evidence suggests that the different animal populations, although tolerant of large changes in salinity, exhibit considerable habitat preference within salt-marshes which results in marked non-random distributions of populations.

THE PRODUCTIVITY OF SALT-MARSHES

As discussed in the introduction, salt-marshes are natural eutrophic systems in which there is usually an abundance of essential elements. The continual cycling of these elements within such an ecosystem is linked to the considerable biological activity which occurs. Temperate salt-marsh ecosystems are often systems of high gross and net productivity with normally an excess of nutrients and no dominant herbivore, and they have stabilised at high levels for both the standing stocks and the fluxes of essential elements and energy

TABLE 3

ANNUAL MEAN NET PRODUCTIVITY

Location	*Principal species*	*Estimated[a] annual net productivity g (dry wt)/m²/year*	*Reference*
Georgia, USA	*Spartina alterniflora*	3 700	Odum (1959) Pomeroy (1959)
North Carolina, USA	*Spartina alterniflora*	650	Williams and Murdoch (1969)
Bridgwater Bay, UK	*Spartina anglica*	960	Ranwell (1961)
Norfolk, UK	*Spartina anglica*	980	Jefferies (unpublished)
Norfolk, UK	*Limonium vulgare*	1 050	Jefferies (unpublished)
Norfolk, UK	*Salicornia* spp.	867	Jefferies (unpublished)

[a] Values based on frequent measurements of total standing crop.

moving through the system (Pomeroy, 1970). Values of the net productivity for *Spartina* marshes may exceed 1 000 g/m^2/year and corresponding values for general salt-marsh communities are also high (Table 3), comparing favourably with values for other ecosystems (Westlake, 1963). The standing crop of perennial plants is augmented by the growth of benthic algae and phytoplankton. Pomeroy (1959) has estimated that algae living on the surface of the mud in a *Spartina* marsh in Georgia contribute substantially to the total net annual primary production and at Stiffkey in Norfolk the standing crop of algae is 25% of the total value at its maximum for a general salt-marsh community (Jefferies, unpublished). The possibility of harvesting the annual crop of *Spartina* for silage has been examined by Hubbard and Ranwell (1967) who fed this material to sheep and concluded that the silage qualities were similar to those of medium quality hay. Although some high standing crop values have been obtained there are instances where the values may be low. For example, in unstable conditions such as occur at the extreme seaward edge of marshes, where the extent of colonisation is limited, the biomass per unit area is low and the vegetation is often composed of a large number of annual plants such as *Salicornia* spp. and *Suaeda maritima*. Likewise in very old marshes or in 'high marshes' the standing crop may be low. Tyler (1971) quotes values as low as 230 g/m^2 for brackish-water meadows at the edge of the Baltic.

One of the most interesting observations relating to the primary productivity of salt-marshes was made by Teal (1962). He estimated that the net primary productivity was only 20% of the gross productivity, the difference reflected energy lost as a result of high respiration rates. Teal suggests that these high rates may be linked with the considerable ionic and osmotic adjustments made by plants in salt-marshes. However, in spite of these high respiration rates the net production is still 1·4% of the incident light energy, a value which compares favourably with other primary producers. It would seem worth while examining whether this large difference between gross and net primary productivity is characteristic of salt-marsh plants in general and to investigate whether it is related to ionic and osmotic adjustments made by these plants.

Besides the primary producers, bacteria and fungi appear to be the most important consumers of energy in salt-marshes (Teal, 1962) and to this extent the complex food chains which are well developed in other ecosystems are reduced in this type of habitat so that the flow of energy is largely short-circuited. However, the action of micro-organisms in breaking down the vegetational litter of salt-marshes is poorly understood. Besides mechanical breakdown associated with tidal movement, Teal has estimated that provided the vegetation was covered by every tide the breakdown and decomposition of the *Spartina* crop would take two months to complete in the case of leaves and three months in the case of stems. Leaves and stems that dry out in periods between spring tides apparently last approximately twice as long. As a result of measurements made throughout the year Teal was able to estimate that the activites of bacteria, averaged over the marsh area for a period of one year, accounted for the breakdown of nearly two-thirds of the available *Spartina* crop. There is, nevertheless, a considerable delay between the death of *Spartina* shoots and their decay—a delay which may limit the food supply of detritus feeders at certain times of the year and hence may affect the abundance of organisms. Dead *Spartina* shoots persist in sheltered ungrazed marshes such as at Poole Harbour, Dorset, for at least five months and possibly up to a year or more before being incorporated in the litter layer (Ranwell, personal communication). However, the turnover of algal material is continuous and the growth of algae is maintained throughout the year. It is of interest in this connection that the species of crab, annelids, nematodes, snails and mussels which are present all have very unrestricted diets and can therefore exploit a range of food supplies available at different seasons. These studies of Teal also emphasise the role of salt-marshes in enabling estuaries to maintain a high productivity. He has shown that particular areas of marsh may export as much as 45% of the *Spartina* crop as a result of tidal action and hence the marsh consumers have no opportunity to utilise this material. He suggests that as the waters in estuaries are often turbid and phytoplankton production is limited (Ragotzkie, 1959), the abundance of animals is largely the result of the export of marsh material, hence events taking place in salt-marshes may subsequently influence the growth of consumers at sites in estuaries removed from the immediate vicinity of marshes.

In addition to the detritus feeders salt-marshes contain a number of herbivorous insects which feed directly on living plants (Odum and Smalley, 1959). Although here again the number of studies is very limited the consumption of energy by this group appears to be 10% of the net primary production, while that of secondary consumers is less than 1% (Teal, 1962). Unfortunately, there appear to be no studies on the energy budget of birds and spiders that feed in salt-marshes. For the purpose of his calculation Teal (1962) assumed that these groups of animals take the same portion of their prey as do the predators of the detritus feeders. However, this may not be valid since some common estuarine birds such as teal (*Anas crecca*) are omnivores feeding on invertebrate animals as well as *Salicornia* (McLusky, 1971).

In a recent paper, Packham and Liddle (1970) suggest that grazing pressure exerted by birds may be considerable, as wildfowl, waders, wood-pigeons, tits and finches feed on some marshes in considerable numbers at different seasons of the year. As a result some minor corrections to patterns of energy flow within salt-marsh communities may be anticipated as more data become available.

ACKNOWLEDGEMENTS

I would particularly like to thank Dr A. J. Gray, Dr G. R. Stewart and Dr G. Tyler who generously sent me unpublished papers and also Dr A. J. Gray, Dr H. Greenway, Dr D. S. Ranwell and Dr A. P. Sims for their valuable criticism of the draft of this paper. Miss S. Sykes kindly typed the manuscript.

REFERENCES

Antonovics, J. (1968). 'Evolution in closely adjacent plant populations.' *Heredity*, **23,** 507–24.

Arisz, W. H., Camphius, I. J., Heikens, H. and Tooren, A. J. van (1955). 'The secretion of the salt glands of *Limonium latifolium* Ktze.' *Acta. Bot. Neerl.*, **4,** 322–38.

Aston, J. L. and Bradshaw, A. D. (1966). 'Evolution in closely adjacent plant populations. II. *Agrostis stolonifera* in maritime habitats.' *Heredity*, **21,** 649–64.

Atkinson, M. R., Findlay, G. P., Hope, A. B., Pitman, M. G., Saddler, H. D. W. and West, K. R. (1967). 'Salt regulation in the mangroves *Rhizophora mucronata Lam.* and *Aegialitis annulata* R. Br.' *Aust. J. Biol. Sci.*, **20,** 589–99.

Baxter, R. M. and Gibbons, N. E. (1954). 'The glycerol dehydrogenases of *Pseudomonas salinaria, Vibrio costicolus* and *Escherichia coli* in relation to bacterial halophilism.' *Canad. J. Biochem. Physiol.*, **32,** 206–26.

Baxter, R. M. and Gibbons, N. E. (1956). 'Effects of sodium and potassium on certain enzymes of *Micrococcus halodenitriticans* and *Pseudomonas salinaria.*' *Canad. J. Microbiol.*, **2,** 599–606.

Baxter, R. M. and Gibbons, N. E. (1957). 'The cysteine desulphydrase of *Pseudomonas salinaria.*' *Canad. J. Microbiol.*, **3,** 461–5.

Bayley, S. T. (1966). 'Composition of ribosomes of an extremely halophilic bacterium.' *J. Mol. Biol.*, **15,** 420–7.

Bayley, S. T. and Kushner, D. J. (1964). 'The ribosomes of the extremely halophilic bacterium, *Halobacterium cutirubrum.*' *J. Mol. Biol.*, **9,** 654–69.

Beevers, L. and Hageman, R. H. (1969). 'Nitrate reduction in higher plants.' *Ann. Rev. Pl. Physiol.*, **20,** 495–522.

Bell, F. G. (1969). 'The occurrence of southern, steppe and halophyte elements in Weichselian (Last-glacial) floras from southern Britain.' *New Phytol.*, **68,** 913–22.

Bernstein, L. (1962). 'Salt-affected soils and plants.' *UNESCO Arid Zone Res.*, **18,** 139–74.

Biebl, R. (1962). In *Physiology and Biochemistry of Algae*. Ed. R. A. Lewin. New York: Academic Press. Pp. 799–815.

Blinks, L. R. (1951). In *Manual of Phycology*. Ed. G. M. Smith, Chronica Botanica, Waltham, Mass., pp. 263–91.

Boorman, L. A. (1967). 'Biological flora of the British Isles. *Limonium vulgare* Mill. and *L. humile* Mill.' *J. Ecol.*, **55,** 221–32.

Boorman, L. A. (1971). 'Studies in salt-marsh ecology with special reference to the genus *Limonium.*' *J. Ecol.*, **59,** 103–20.

Boorman, L. A. and Woodell, S. R. J. (1966). 'The topograph—an instrument for measuring microtopography.' *Ecology*, **47,** 869–70.

Brereton, A. J. (1971). 'The structure of the species populations in the initial stages of salt-marsh succession.' *J. Ecol.*, **59,** 321–38.

Cajander, A. K. (1903). 'Beiträge zur Kenntnis der Vegetation der Alluvionem des nördlichen Eurasiens 1.' *Acta. Soc. Sci. Fenn.*, **32,** 1.

Chapman, V. J. (1960). *Salt-Marshes and Salt Deserts of the World.* London and New York: Leonard Hill/Interscience.

Christian, J. H. B. and Waltho, J. A. (1961). 'Solute concentrations within cells of halophilic and non-halophilic bacteria.' *Biochem. Biophys. Acta*, **65,** 506–8.

Dalby, D. H. (1970). 'The salt-marshes of Milford Haven, Pembrokeshire.' *Field Studies*, **3,** 297–310.

Duffey, E. (1970). 'Habitat selection by spiders on salt marsh in Gower.' *Nature in Wales*, **12,** 15–23.

Epstein, E. (1969). 'Mineral metabolism of halophytes.' In *Ecological Aspects of the Mineral Nutrition of Plants.* Ed. I. H. Rorison (British Ecological Society Symposium.) Oxford: Blackwell.

Evans, H. J. and Sorger, G. J. (1966). 'Role of mineral elements with emphasis on the univalent cations.' *Ann. Rev. Pl. Physiol.*, **17,** 47–71.

Gimingham, C. H. (1964). 'Maritime and sub-maritime communities.' In *The Vegetation of Scotland.* Ed. J. H. Burnett. Edinburgh: Oliver and Boyd.

Goodman, P. J. (1960). 'Investigations into "die-back" in *Spartina townsendii* agg. II. The morphological structure and composition of the Lymington sward.' *J. Ecol.*, **48,** 711–24.

Gray, A. J. (1971). *Variation in Aster tripolium, L. with particular reference to some British populations*. Ph.D. Dissertation, University of Keele.

Gray, A. J. and Bunce, R. G. H. (1972). 'The salt-marshes of Morecambe Bay. II. Soils and Vegetation—a multivariate approach.' *J. App. Ecol.* (in press).

Greenway, H. and Pitman, M. G. (1965). 'Potassium retranslocation in seedlings of *Hordeum vulgare.*' *Aust. J. Biol. Sci.*, **18,** 235–47.

Greenway, H. and Thomas, D. A. (1965). 'Plant response to saline substrates. V. Chloride regulation in individual organs of *Hordeum vulgare* during treatment with sodium chloride.' *Aust. J. Biol. Sci.*, **18,** 505–24.

Greenway, H., Gunn, A. and Thomas, D. A. (1966). 'Plant response to saline substrates. VIII. Regulation of ion concentrations in salt-sensitive and halophytic species.' *Aust. J. Biol. Sci.*, **19,** 741–56.

Gregor, J. W. (1944). 'The ecotype.' *Bio. Rev.*, **19,** 20–30.

Gutknecht, J. and Dainty, J. (1969). 'Ionic relationships of marine algae.' *Oceanogr. Mar. Biol. Ann. Rev.*, **6,** 163–200.

Harper, J. L. (1967). 'A Darwinian approach to plant ecology.' *J. Ecol.*, **55,** 247–71.

Hayward, H. E. and Wadleigh, C. H. (1949). 'Plant growth on saline and alkaline soils.' *Adv. Agron.*, **1,** 1–38.

Hayward, H. E. and Bernstein, L. (1958). 'Plant-growth relationships on salt-affected soils.' *Bot. Rev.*, **24,** 584–635.

Hill, A. E. (1967a). 'Ion and water transport in *Limonium.* I. Active transport by leaf gland cells.' *Biochim. Biophys. Acta,* **135,** 454–60.

Hill, A. E. (1967b). 'Ion and water transport in *Limonium.* II. Short-circuit analysis.' *Biochim. Biophys. Acta*, **135,** 461–5.

Hill, A. E. (1970a). 'Ion and water transport in *Limonium.* III. Time constants of the transport system.' *Biochim. Biophys. Acta*, **196,** 66–72.

Hill, A. E. (1970b). 'Ion and water transport in *Limonium.* IV. Delay effects in the transport pumps.' *Biochim. Biophys. Acta*, **196,** 73–9.

Holmes, P. K. and Halvorson, H. O. (1965). 'Purification of a salt-requiring enzyme from our obligately halophilic bacterium.' *J. Bact.*, **90,** 312–5.

Hubbard, J. C. E. and Ranwell, D. S. (1967). 'Cropping *Spartina* salt marsh for silage.' *J. Br. Grassld Soc.*, **21,** 214–7.

Jennings, D. H. (1968). 'Halophytes, succulence and sodium in plants—a unified theory.' *New Phytol.*, **67,** 899–911.

Larsen, H. (1967). 'Biochemical aspects of extreme halophilism.' *Adv. Micro. Physiol.*, **1,** 97–132.

Lubin, M. (1964). 'Cell potassium and the regulation of protein synthesis.' In *The Cellular Functions of Membrane Transport*. Ed. J. F. Hoffman. New Jersey: Prentice-Hall.

Lüttge, U. (1971). 'Structure and function of plant glands.' *Ann. Rev. Pl. Physiol.*, **22,** 23–44.

Luxton, M. (1964). 'Some aspects of the biology of salt-marsh Acarina.' *Acarologia*. (C.R.1er Congress Int. d'Acarologie, Fort Collins, Col., USA, 1963). Pp. 172–82.

MacDonald, K. B. (1969). 'Quantitative studies of salt marsh mollusc faunas from the North American Pacific Coast.' *Ecol. Monog.*, **39,** 33–59.

MacRobbie, E. A. C. (1971). 'Fluxes and compartmentation in plant cells.' *Ann. Rev. Pl. Physiol.*, **22,** 75–96.

McLusky, D. S. (1971). *Ecology of Estuaries*. London: Heinemann.

Monoszon, M. C. (1964). 'Pollen of halophytes and zerophytes of the Chenopodiaceae family in the periglacial zone of the Russian plain.' *Pollen Spores*, **6,** 147.

Odum, E. P. (1959). *Fundamentals of Ecology*. 2nd ed. Philadelphia: Saunders.
Odum, E. P. and Smalley, A. E. (1959). 'Comparison of population energy flow of a herbivorous and deposit-feeding invertebrate in a salt-marsh ecosystem. *Proc. Nat. Acad. Sci., USA*, **45,** 617–22.
Osmond, C. B. (1970). 'Carbon metabolism in *Atriplex* leaves.' In *Biology of Atriplex*. Ed. R. Jones. Canberra: CSIRO.
Packham, J. R. and Liddle, M. J. (1970). 'The Cefni salt-marsh, Anglesey, and its recent development.' *Field Studies*, **3,** 311–56.
Parham, M. R. (1971). *A comparative study of the mineral nutrition of selected halophytes and glycophytes*. Ph.D. Dissertation, University of East Anglia.
Perring, F. H. and Walters, S. M. (1962). *Atlas of the British Flora*. London: Nelson.
Pigott, C. D. (1969). 'Influence of mineral nutrition on the zonation of flowering plants in coastal salt-marshes.' In *Ecological Aspects of the Mineral Nutrition of Plants*. Ed. I. H. Rorison. (British Ecological Society Symposium). Oxford: Blackwell.
Pomeroy, L. R. (1959). 'Algal productivity in the salt-marshes of Georgia.' *Limnol. Oceanogr.*, **4,** 386–97.
Pomeroy, L. R. (1970). 'The strategy of mineral cycling.' *Ann. Rev. Ecol. & Systems*, **1,** 171–90.
Pomeroy, L. R., Johannes, R. E., Odum, E. P. and Roffman, B. (1969). 'The phosphorus and zinc cycles and productivity of a salt-marsh.' In *Proc. 2nd Symp. Radioecol. US At. Energy Comm. TID* 4500. Ed. D. J. Nelson and F. C. Evans. Pp. 412–9.
Pritchard, D. W. (1967). 'What is an estuary: Physical view point.' In *Estuaries*. Ed. G. H. Lauff. Publication No. 83. American Association for the Advancement of Science, Washington, D.C.
Ragotzkie, R. A. (1959). 'Plankton productivity in estuarine waters of Georgia.' *Inst. Marine Sci.*, **6,** 146–58.
Raju, P. V. and Jefferies, R. L. (1972, in preparation). 'The effect of inorganic cations on the activity of enzymes of halophytic plants.'
Ranwell, D. S. (1961). '*Spartina* salt-marshes in southern England. I. The effects of sheep grazing at the upper limits of *Spartina* marsh in Bridgwater Bay.' *J. Ecol.*, **49,** 325–40.
Ranwell, D. S. (1968). *Coastal Marshes in Perspective*. Regional Studies Group Bulletin, No. 9, University of Strathclyde.
Ranwell, D. S. and Downing, B. M. (1959). 'Brent goose (*Branta bernicla* (L.)) winter feeding pattern and *Zostera* resources at Scolt-Head Island, Norfolk.' *Animal Behaviour*, **7,** 42–56.
Ranwell, D. S., Bird, E. C. F., Hubbard, J. C. E. and Stebbings, R. E. (1964). '*Spartina* salt marshes in southern England. V. Tidal submergence and chlorinity in Poole Harbour.' *J. Ecol.*, **52,** 627–41.
Ryther, J. H. and Dunstan, W. M. (1971). 'Nitrogen, phosphorus and eutrophication in the coastal marine environment.' *Science*, **171,** 1008–13.
Scholander, P. F., Hammel, H. T., Hemmingsen, E. A. and Garey, W.

(1962). 'Salt balance in mangroves.' *Pl. Physiol., Lancaster*, **37,** 722–9.

Scholander, P. F., Bradstreet, E. D., Hammel, H. T. and Hemmingsen, E. A. (1966). 'Sap concentrations in halophytes and some other plants.' *Pl. Physiol., Lancaster*, **41,** 529–32.

Sharrock, J. T. R. (1967). *A study of morphological variation in Halimione portulacoides (L.) Aell. in relation to variations in the habitat*. Ph.D. Dissertation, University of Southampton.

Sims, A. P., Stewart, G. R. and Folkes, B. F. (1972, in preparation). 'The effect of inorganic cations on the activity of plant enzymes; differences in enzyme properties in relation to species and habitat.

Slatyer, R. O. (1967). *Plant-Water Relationships*. London: Academic Press.

Stebbings, R. E. (1971). 'Some ecological observations on the fauna in a tidal marsh to woodland transition.' *Proc. Trans. Brit. Ent. Soc.*, **4,** 83–8.

Stewart, G. R., Lee, J. A. and Orebamjo, T. O. (1972). 'Nitrogen metabolism of halophytes. I. Nitrate reductase activity in *Suaeda maritima*.' *New Phytol.* (in press).

Strogonov, B. P. (1964). *Physiological basis of salt tolerance of plants*. London: Oldbourne.

Taylor, M. C. and Burrows, E. M. (1968). 'Studies on the biology of *Spartina* in the Dee Estuary, Cheshire.' *J. Ecol.*, **56,** 795–809.

Teal, J. M. (1962). 'Energy flow in the salt-marsh ecosystem of Georgia.' *Ecology*, **43,** 614–24.

Tyler, G. (1967). 'On the effect of phosphorus and nitrogen, supplied to Baltic shore-meadow vegetation.' *Bot. Notiser*, **120,** 433–47.

Tyler, G. (1971). 'Distribution and turnover of organic matter and minerals in a shore meadow ecosystem. Studies in the ecology of Baltic sea shore meadows IV.' *Oikos*, **22** (in press).

Westlake, D. F. (1963). 'Comparisons of plant productivity.' *Bio. Rev.*, **38,** 385–425.

Williams, R. B. and Murdoch, M. B. (1969). 'The potential importance of *Spartina alterniflora* in conveying zinc, manganese and iron into estuarine food chains.' In *Proc. 2nd Symp. Radioecol. US At. Energy Comm. TID* 4500. Ed. D. J. Nelson and F. C. Evans. Pp. 431–9.

Wit, C. T. de (1960). 'On competition.' *Verslland bouwk. Onderz. Ned.*, **66,** 8.

Woodell, S. R. J. and Mooney, H. A. (1970). 'The effect of sea water on carbon dioxide exchange by the Halophyte *Limonium californicum* (Boiss) Heller.' *Ann. Bot.*, **34,** 117–21.

6

Standing Crop, Productivity and Trophic Relations of the Fauna of the Ythan Estuary

H. MILNE and G. M. DUNNET

INTRODUCTION

Work currently being carried out on the Ythan estuary by members of staff, research students and undergraduates of the Zoology Department, University of Aberdeen, is part of a long-term programme planned in 1964. The aim is to quantify the seasonably variable exploitation of the invertebrate fauna of the estuary by the wide variety of predators present, mainly birds and fish. In the past seven years, three staff members, ten Ph.D. students and about ten undergraduates have participated actively in research projects. Some of the results have already been published but many data are still in unpublished theses. This paper is the first attempted synthesis of the work. Whereas we are primarily concerned here with the interrelations between populations of animals within the estuary, this particular estuarine ecosystem also offered the opportunity to make detailed population studies *per se*, in a relatively natural situation.

Our studies have been deliberately confined to the upper trophic levels of the ecosystem; we believe that a useful contribution can be made to the understanding of estuarine food chains, and to predator-prey interrelations in particular, without becoming involved, initially at least, in studies of the food and feeding of the prey. Bird predators are easily observed and quantitative data on their numbers, distribution, feeding behaviour, feeding rates, and diet can readily be obtained, but the fish species are much more difficult to study in this context, and we have to rely on various catching methods to provide much of the data. The invertebrate prey species are readily accessible, in intertidal areas exposed at low tide, allowing studies of distribution, dispersion pattern, standing crop numbers and biomass, and productivity to be made.

The project is recognised as part of the UK's supporting programme to the International Biological Programme (PM Section).

THE STUDY AREA

The Ythan estuary, opening into the North Sea 20 km north of Aberdeen is about 8 km long and averages 300 m wide throughout its length. The total area is about 240 ha of which 165 ha are intertidal at spring tides. The tidal influence is very strong, with a tidal volume to fresh-water volume ratio of 20:1 at high water, and complete flushing of the estuary occurs between tides (Leach, 1971). There is a marked seasonality in the amounts of dissolved nutrients in the water: levels of nitrates have a peak in early spring following application of agricultural fertiliser to fields within the drainage area of the river, and silicates also are very largely determined by the fresh water input, whereas levels of phosphates parallel those of the local sea water (Leach, *loc. cit.*).

The variety of substrata in the intertidal regions of the estuary ranges from mud and mud/sand at the head of the estuary, through mud/sand and shingle (with attached mussels) in the middle reaches, to pebbles, sand and boulders (with mussels) at the mouth (Anderson, 1971, Hinton and Milne, unpublished). Over 90% of the total area of the estuary consists of mud or mud/sand (200 ha) and mussel beds (30 ha), and this can conveniently be divided into two major communities—the mudflats and the mussel beds—as a basis for study.

FOOD WEB

The invertebrates are more or less confined to specific substrata according to their requirements. Factors such as water current, particle size of the substratum, exposure and salinity also play a role in limiting the variety of species which occur and determine their distribution within the estuary. Sixty-three species of invertebrates have been recorded (Table 1 and Appendix). The predators are highly mobile and exhibit obvious tidal rhythms of activity associated with the activity of their prey and water movements. Some birds feed mainly on the intertidal zone at low water (*e.g.* Redshank or Shelduck), some feed in the shallow water at low tide (*e.g.* Eiders), some feed at the edge of the tide as it ebbs and flows (*e.g.* Shelducks, gulls) and others feed at high water by diving (diving ducks, Cormorants, Mergansers) and plunge diving (terns). In contrast the fish feed actively within the intertidal zone only when these areas are

TABLE 1

SPECIES COMPOSITION OF THE YTHAN ESTUARY FAUNA

	Groups	*No. of spp.*	*Totals*
Predators	Birds	53	75
	Fish	22	
Prey	Molluscs	12	63
	Crustaceans	26	
	Annelids	15	
	Other inverts.	c. 10	

covered by water (Flounders and Gobies) or they feed within the water column (Sea Trout). Such patterns of activity dictate not only when and for how long any particular predator may feed, but also the prey upon which they may feed; and activity patterns of prey species in relation to tidal ebb and flow influence their availability to predators.

In the mudflat communities *Hydrobia ulvae, Corophium volutator, Nereis diversicolor* and *Macoma balthica* predominate in the diets of the various predators (Fig. 1) while, on the mussel beds, *Mytilus edulis*, gammarids, and *Carcinus maenas* are the major food species (Fig. 2). The food webs shown in Figs. 1 and 2 have been simplified

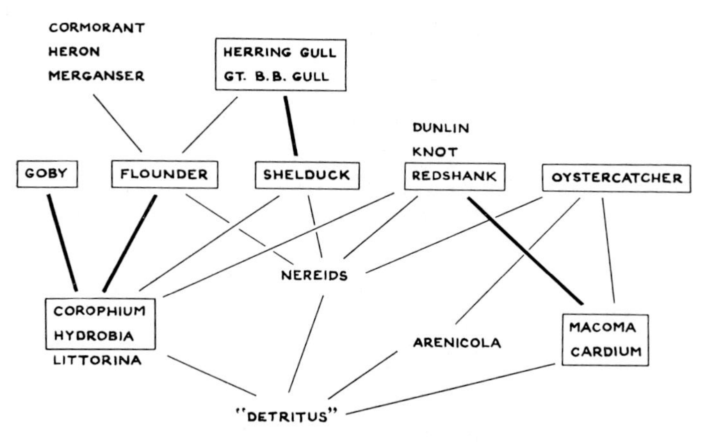

FIG. 1. Food web of mudflat community.

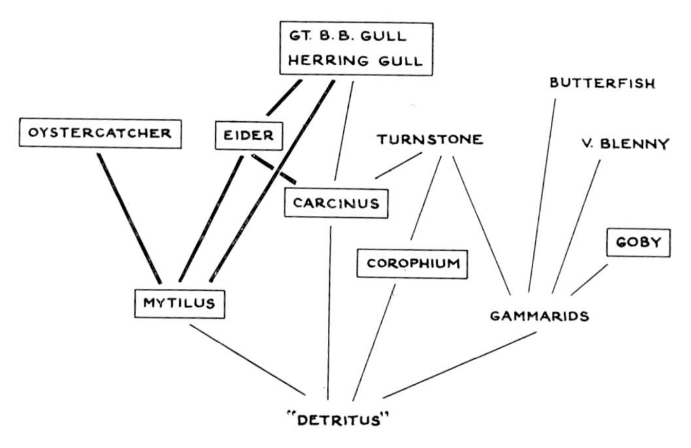

FIG. 2. Food web of mussel bed community.

for clarity; the boxes and thicker lines indicate where studies have already taken place or are in progress. These may be either purely qualitative, describing diets of predators or the presence/absence of prey in certain localities, or they may be quantitative to the extent that standing crop numbers (or densities), biomass, and productivity have been measured (boxed names). Thin lines in the diagrams indicate that species A eats species B, whereas thick lines denote where the amounts eaten by species A, or the energy flow, have been measured.

THE PREDATORS

All the predators exhibit marked seasonal changes in numbers in the estuary. Birds using the estuary fall into four categories (i) summer (breeding) visitors, (ii) winter visitors, (iii) passage migrants and (iv) residents throughout the year. The fish may be mainly summer visitors, such as Flounders, or autumn/winter visitors, such as the Gobies, fattening between breeding seasons. Sea Trout, on the other hand, are passage migrants with maximum numbers in summer.

Mudflat community

Twenty-one species of wading birds have been recorded in the estuary, but only eight occur regularly and in significant numbers, *i.e.* Redshank, Dunlin, Knot, Curlew, Golden Plover, Lapwing,

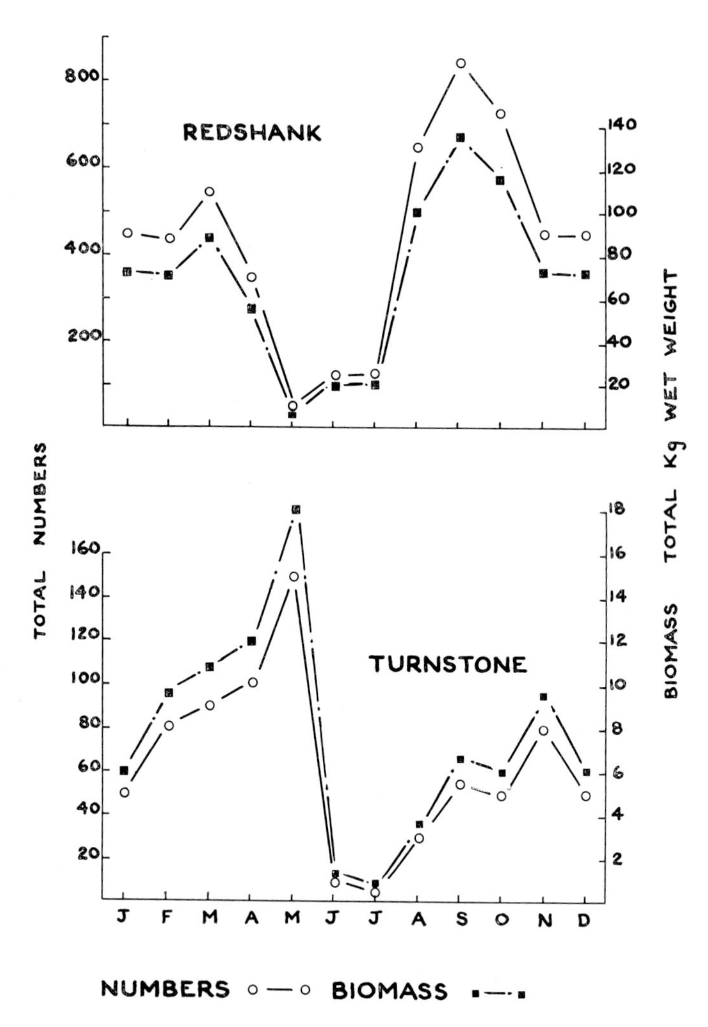

FIG. 3. Seasonal changes in numbers and biomass of Redshank and Turnstone in the Ythan estuary.

Oystercatcher and Turnstone. Most of these appear for relatively short periods of time, and, as examples, the seasonal pattern of total numbers and biomass of Redshank and Turnstone is shown in Fig. 3 (data from Goss-Custard, 1966, 1969, 1970).

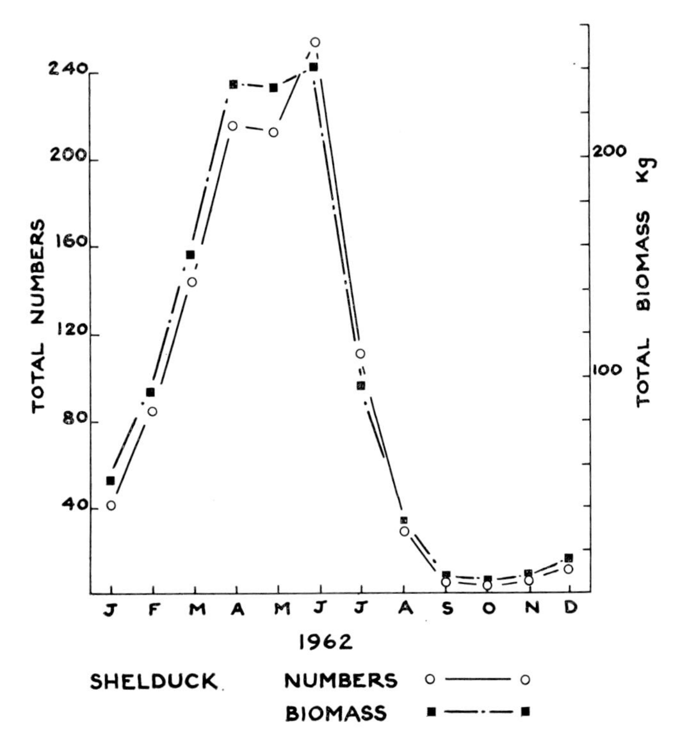

FIG. 4. Seasonal changes in numbers and biomass of Shelduck in the Ythan estuary.

Most waders concentrate on the mudflats when feeding, but Oystercatchers, Turnstones and Redshank use the mussel beds extensively as well.

Of the ten species of waterfowl regularly seen on the estuary the Shelduck are typically associated with mudflats, where they hold feeding territories on the intertidal areas from early spring to mid-summer (Young, 1964, 1970). These birds, although relatively few in numbers and sparsely dispersed over the mudflats, reach a fairly high total standing crop biomass in summer (Fig. 4) and their food intake from the mudflat community is large because of their relatively high metabolic rate.

Flounders and Gobies may reach very substantial densities (Healey, 1971) and standing crop biomass in summer (Figs. 5, 6). Estimates indicate that the standing crop biomass for the whole estuary fluctuates seasonally between 5 and 20 metric tonnes fresh weight

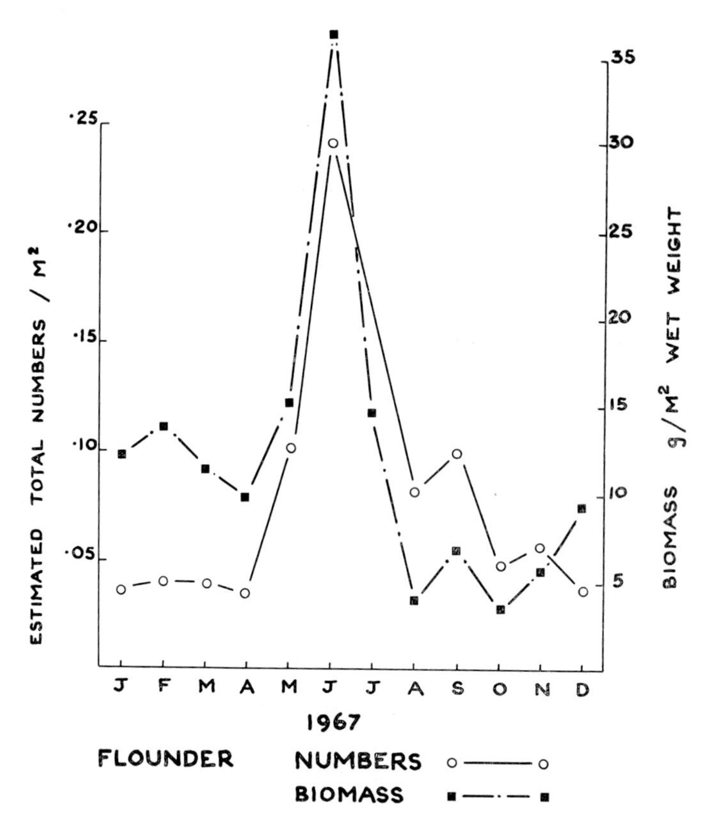

FIG. 5. Seasonal changes in numbers and biomass of Flounders in the Ythan estuary.

for Gobies and 10 to 100 metric tonnes for Flounders. Their metabolic requirements, per unit weight, however, are considerably lower than those of birds and their overall consumption of food from the system may be of a similar order of magnitude to that of the birds.

Mussel bed community

Birds and man are the major predators on mussels. Eiders, which show a marked seasonal change in numbers (Milne, 1963), represent a total standing crop fresh weight biomass fluctuating seasonally betweeen 1 and 5 metric tonnes (Fig. 7). Not only does their total biomass change through the year but also their food requirements and diet change from summer to winter. In winter they have a relatively high consumption rate (*c*. 600 K cals/bird/day) composed

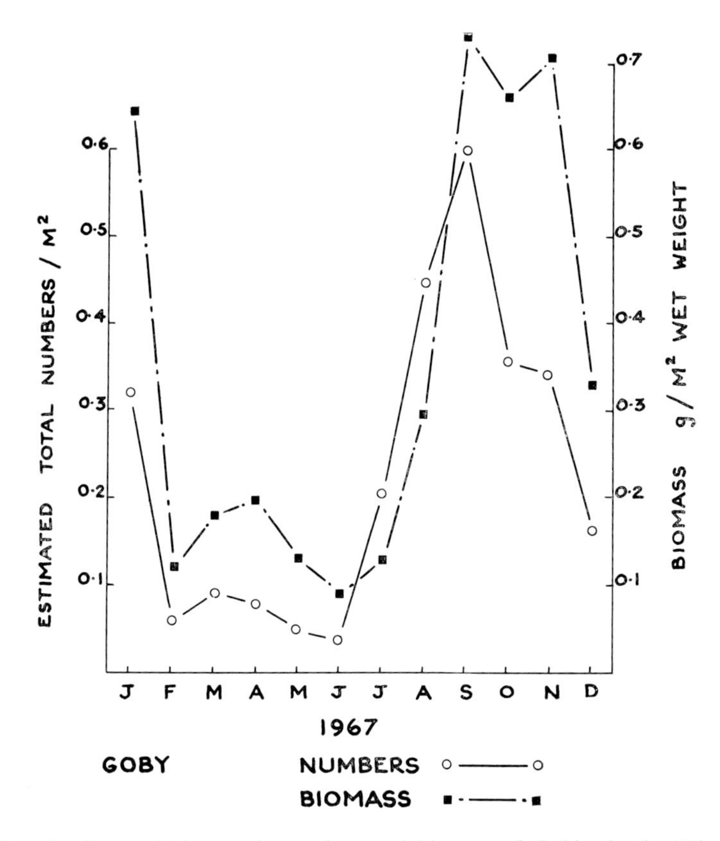

FIG. 6. Seasonal changes in numbers and biomass of Gobies in the Ythan estuary.

almost entirely of mussels (Marriott, 1966). In spring the females, when fattening for breeding, double their consumption rate for a period of about four weeks and then stop feeding completely for four weeks during incubation of the eggs (Gorman and Milne, 1971). *Carcinus maenas* and *Littorina littorea* form a major part of the summer diet thus reducing the demand on mussels at the time of peak numbers of Eiders. Over the year these birds consume about

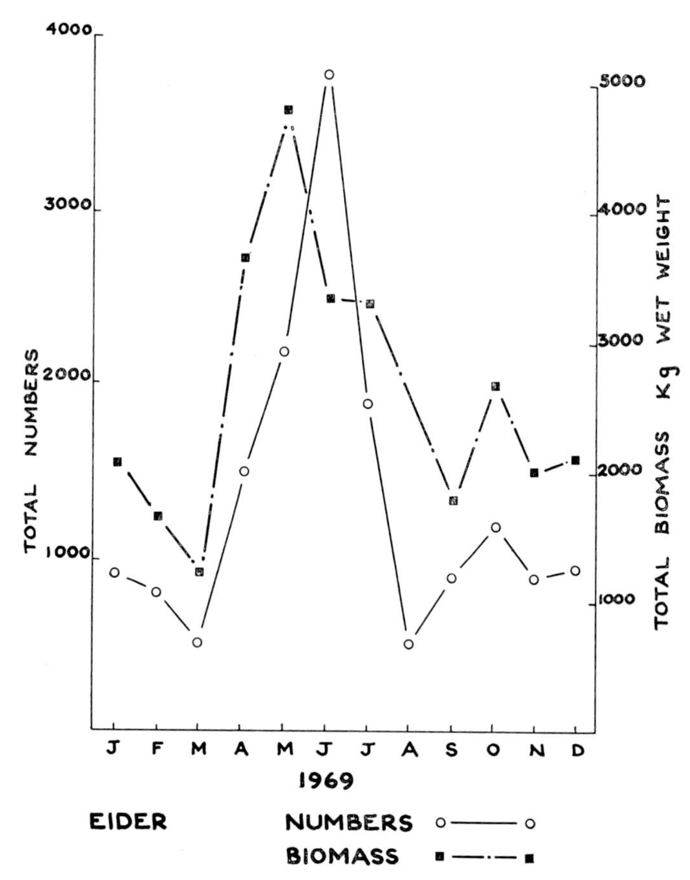

FIG. 7. Seasonal changes in numbers and biomass of Eiders in the Ythan estuary.

20% of the total production, or 39% of the net annual production of *Mytilus* (Milne, 1968). Other main predators on *Mytilus* are Herring Gulls and Oystercatchers (Fig. 2).

Most of the intertidal fishes found in the mussel bed community (*e.g.* Blennies, Butterfish) eat gammarids but, to date, no quantitative studies have been made on their feeding ecology.

THE PREY

As with the predators, there are a large number of prey species (63) recorded from the estuary but only a few (10) are present in high densities (*Hydrobia ulvae, Corophium volutator, Macoma balthica, Nereis diversicolor, Littorina littorea, Cardium edule, Marinogammarus marinus, Mytilus edulis, Arenicola marina, Carcinus maenas*), whilst some are abundant only for very short periods of time (*e.g. Neomysis integer*).

Whereas the predators show seasonal fluctuations in numbers and biomass as a result of migration, the prey species exhibit similar fluctuations as a result of reproduction and mortality. Practically all

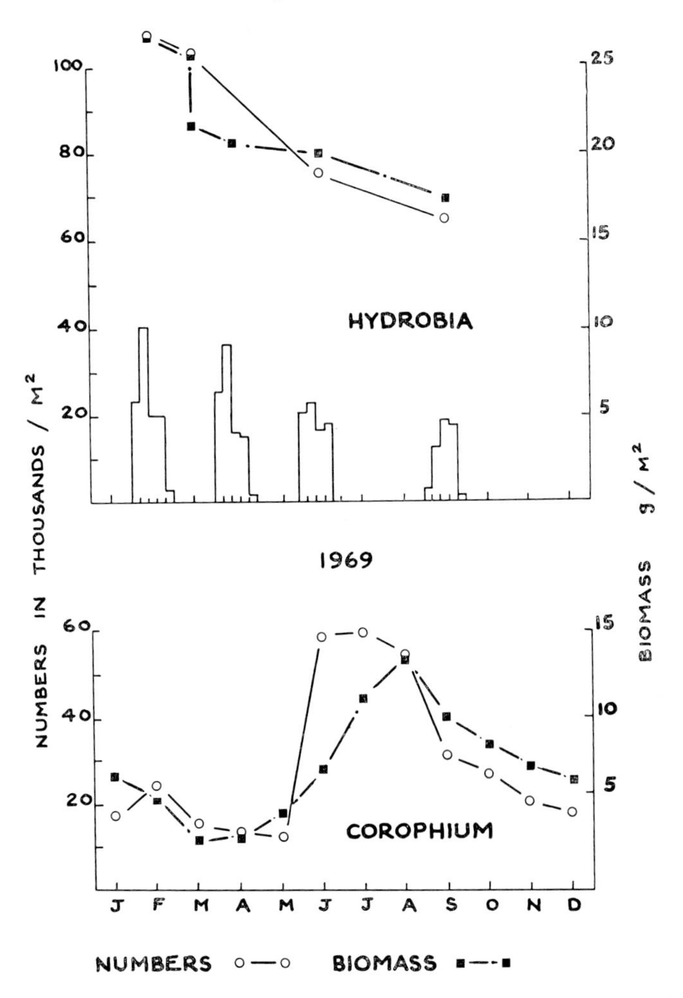

FIG. 8. Seasonal changes in numbers and biomass of *Hydrobia* and *Corophium* in the Ythan estuary.

the species show a general pattern of summer growth and reproduction, producing a peak of numbers and biomass in autumn, followed by a period of winter/spring mortality (and 'negative growth') leading to minimum numbers and biomass the following spring. This pattern is typified by *Hydrobia* (from Anderson, 1971), *Corophium* (Wood, unpublished) and *Mytilus* (Milne and Hinton, unpublished) (Figs. 8, 9).

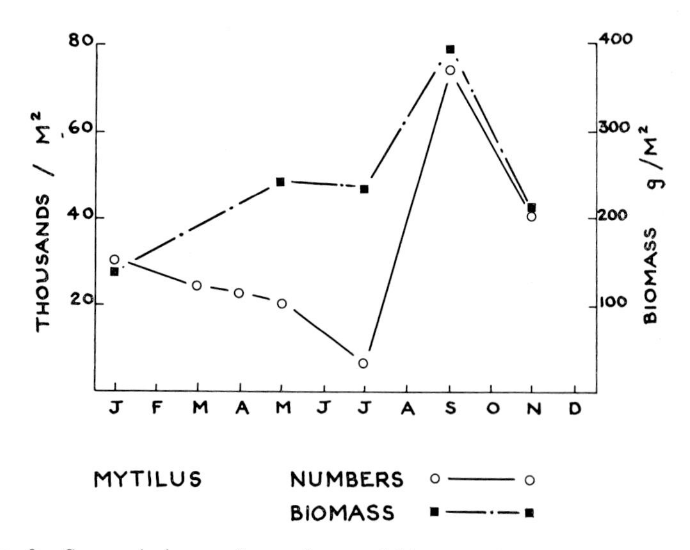

FIG. 9. Seasonal changes in numbers and biomass of *Mytilus* in the Ythan estuary.

Clearly, to support such a large biomass of predators, the prey must maintain a high level of productivity. Studies of the productivity of *Mytilus, Corophium* and *Hydrobia* have been carried out in selected parts of the estuary, and current studies include *Macoma* and *Nereis*. As an example of these productivity estimates the work on *Mytilus* is presented in Table 2.

Productivity of *Mytilus*

The age of different sizes of individual mussels was determined by allocating year classes to the size-frequency modes within the population which could be described on each sampling occasion. This method gave an indication of the age composition of the population. Three age-cohorts were recognised—the spatfall of the previous year, the spatfall of two years previously and a composite group of larger mussels containing individuals from three to ten years old.

TABLE 2

ESTIMATED GROSS PRODUCTION OF *MYTILUS* DURING 1969 AT THE LOW WATER SAMPLING STATION, RED INCHES, YTHAN ESTUARY

A. *Production of survivors*

	Time period	*Mean No. per m^2*	*Individual mean dry ΔW (mg)*	*Production per m^2 (g, dry wt)*
(a) 1968 Cohort	March–May	9 709·14	0·10	0·971
	June–July	6 153·34	1·71	10·522
	August–Sept.	2 466·00	33·68	83·055
			Cohort Total P =	94·548
(b) 1967 Cohort	March–May	433·34	6·71	2·976
	June–July	133·34	80·01	10·669
	August–Sept.	83·34	117·70	9·809
			Cohort Total P =	23·454
(c) 'Large/old group'	March–Sept.	153·32	656·30	100·624 g, dry wt

B. *Production of animals dying during period*

(P = N′ × $\frac{1}{2}\Delta$W of survivors)

		$\overline{N}$	$\frac{1}{2}\Delta W$ *(mg)*	*Production per m^2 (g, dry wt)*
(a) 1968 Cohort	March–May	9 023·88	0·05	0·451
	June–July	3 555·80	0·86	3·040
	August–Sept.	3 687·34	6·74[a]	24·838[a]
				28·329
(b) 1967 Cohort	March–May	1 741·14	3·36	5·843
	June–July	310·00	40·01	12·402
	August–Sept.	50·00	58·85	2·943
				21·188

(c) No detectable mortality in 'Large/old group', March–Sept.

C. *Total Production in* 1969 = 268·153 g, dry wt/m^2
c.v. of Mytilus *flesh* = 5 K cals/g
Therefore P = 1 340·8 K cals/m^2/year

[a] Growth curve of 1968 cohort indicates that only one-fifth of the total weight increment for the period July–September would have been achieved during the first half of that time period.

Growth rates were obtained (*a*) by measuring the change in mean size of individuals from any given cohort of the population and (*b*) by keeping mussels of known size in a cage in the estuary and measuring these at intervals throughout the year.

Estimates of the production of those individuals which survived through the (summer) growing period were obtained by multiplying the numbers surviving by their mean weight increment during the summer. Production estimates for those individuals which were removed by predators during the growth period, but which would have added some weight increment before dying, were obtained by multiplying the numbers which were lost by half the growth increment experienced by the survivors over the same time interval.
i.e.

$$\text{Production of survivors} = N \times \Delta W$$
$$\text{Production of those dying} = N' \times \tfrac{1}{2}\Delta W$$

where N = numbers surviving through growth period, N' = numbers removed during the growth period, and ΔW = weight increment during the growth period.

Calculation of gross production* of *Mytilus* on Red Inches mussel bed during the summer of 1969 is given in Table 2, and gives a total value of 1 300 K cals/m^2 for the period April to September.

Partition of *Mytilus* production

Although three species of birds are the main predators feeding on the mussel beds, their behavioural and morphological adaptations clearly separate their feeding in both time and space. All three make big demands on the mussel population through the winter months, October–April, and all feed mainly at low water. Eiders feed by dabbling in shallow water at the edge of the ebbing tide, or by diving in deeper sub-littoral water and as the tide begins to flood. The mussels they pick up are swallowed whole and are crushed by the birds' large, muscular gizzards; crushed shells are passed through the gut and voided in the faeces. Such a feeding method precludes the intake of mussels larger than about 40 mm in length, and mussels less than 5 mm long were seldom taken. The mean length of mussels eaten by Eiders was 18 mm.

Oystercatchers feed mainly when the mussel beds are completely exposed at low tide. They are highly specialised feeders, their bill being adapted to breaking open mussel shells by hammering on one of the valves until it splits open or by being inserted between the

* This term is used since part of the total weight increment is later utilised in metabolism during the winter and should not, therefore, be included in any value for net annual production.

valves to cut through the adductor muscles (Norton-Griffiths, 1967; Heppleston, 1971). Once the shell is opened the flesh is scooped out and eaten. Size selection once again operates, and mussels between 18 mm and 70 mm (mean of 33 mm) were taken.

Herring Gulls also feed on mussels which are exposed at low tide, eating the very small individuals (2–10 mm long) which are swallowed whole, or picking up and dropping large individuals (greater than 35 mm in length) to break them on the shore. Figure 10 gives a schematic presentation of the partitioning of the summer production of mussel flesh among the various predators.

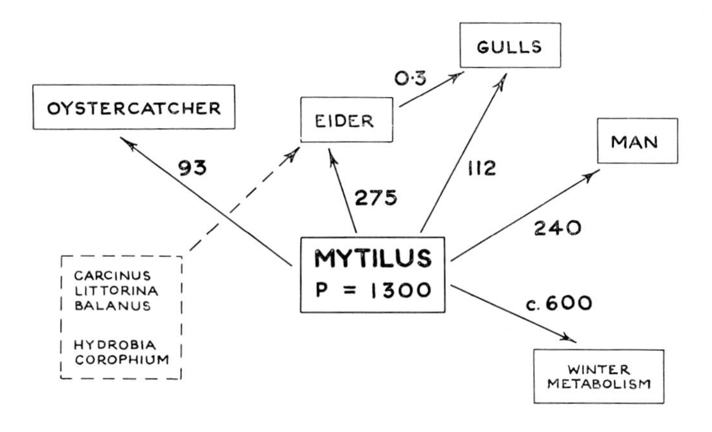

FIG. 10. The partitioning of *Mytilus* production among the various predators of the mussel bed community.

The data are incomplete but the outstanding feature is that virtually all the gross production is accounted for in terms of predation and overwintering metabolic requirements. This suggests that the mussel bed is being cropped to a maximum, so that the standing crop at the beginning of each year remains fairly constant and no net changes in size of the mussel population are occurring.

SUMMARY AND CONCLUSIONS

This paper describes in very general terms, the estuary of the river Ythan in Aberdeenshire, its topography, substrata and hydrology, and outlines, giving selected detailed examples, the qualitative and quantitative interrelations of the fauna of the upper trophic levels of the community. Primary production and the source and rate of deposition of detritus, on which the whole biological community depends, have not been considered. The Ythan estuary is not grossly affected by man: run-off of agricultural chemicals and the input of small amounts of domestic sewage are the main contaminants, but the flushing period of one tidal cycle prevents the accumulation of these materials. Relative to many other estuaries, often much larger and surrounded by industry and suburban development, the Ythan can be considered unpolluted and 'natural'.

The most conspicuous feature of the fauna of the estuary is the high level of flux—in daily and tidal cycles of feeding and movement; in seasonal patterns of abundance, different for each of the many species, especially the highly mobile predators; and in energy flow due to the marked seasonal patterns of growth, reproduction and predation. Despite this high level of flux, the large populations of a few highly productive species of invertebrates support, year after year, up to 75 species of predating birds and fish, each with its own characteristic pattern and level of seasonal abundance, feeding behaviour, diet, and food requirements. The system calls to mind a transit camp, with a well-organised catering department designed to supply the needs of a fluctuating and very diverse clientele, and adaptable enough for the establishment to stay in business! The long-term stability of the system and its natural fluctuations are impossible to demonstrate owing to lack of quantitative information, but mussels have been exploited here by man since the beginning of the Christian era (Walton, 1966). It is of theoretical and practical interest to understand both the qualitative and quantitative aspects of species diversity and interdependence in relation to the apparent long-term stability of the estuarine community as a whole.

Apart from the trophic (catering) aspects of the estuarine community, there is also an accommodation aspect. Many species of birds use the estuary itself, its sand bars, mudflats and shingle beds, for roosting. Several thousand gulls (up to five species), geese (five species), ducks and swans (especially Whooper Swans), and some waders come to the estuary to roost, but may obtain most of their food either at sea or from the agricultural hinterland.

The 'nutrient trap' effect, characteristic of estuaries (Odum, 1961; 1971, p. 357), resulting in a proliferation of filter or deposit-feeding invertebrates is the basis of the high productivity of estuarine systems. The tidal activity, with its accompanying deposition of organic material, effectively renders these animal populations independent of the primary production of their local communities, and allows them to sustain higher standing crops and productivity than they could otherwise achieve. Such levels of productivity, illustrated by the *Mytilus* example, are frequently an order of magnitude higher than the productivity achieved by invertebrate populations in terrestrial or freshwater communities (McNeill and Lawton, 1970; Petrusewicz, 1967; Engelmann, 1966).

It is the combination of this variety and productivity of detritus and filter feeders within any habitat, and the complexity of habitats within the estuary, which sets the stage for the diversity of predators in the system as a whole.

In problems relating to the management of estuaries, particularly where loss of habitat is involved, the question of within-habitat diversity is important in considering which habitats may be reduced without an accompanying decrease in overall species variety. Also, the manner in which the prey production is partitioned among predators is important if one is to consider whether a loss of prey habitat (and hence of food for the predators) is likely to result in a decline in the total number of predators. In the situation where food is superabundant, a partial reduction in the supply of that resource need not cause any change in the numbers of predators, but attention must be paid to the seasonal pattern of abundance, and the distribution in relation to feeding habitats, of the various predator species.

ACKNOWLEDGEMENTS

Much of the work described here is, as yet, unpublished and we are grateful to the following for access to their data:

Dr J. Hinton for data on *Mytilus* numbers and productivity; A. Anderson for *Hydrobia* numbers and his faunal list; T. Wood for data on *Corophium* numbers and biomass, and *Hydrobia* weights; D. C. Emerson for Flounder densities; J. A. Love for data on predation on mussels by Herring Gulls, Oystercatchers and Man; R. Summers for weights of Flounders and waders and M. Williams for Shelduck weights. We are also indebted to Dr W. J. Bates of the Udny Arms Hotel, Newburgh, and the Ythan River Board for data on the mussel harvest, for permission to sample the mussel beds and to use a beach seine net to sample Flounders on the estuary throughout the year.

Mr A. Anderson very kindly made all the text figure drawings.

During the study financial support has come from NERC Research Grants and Studentships, and from the University of Aberdeen.

REFERENCES

Anderson, A. (1971). 'Intertidal activity, breeding and the floating habit of *Hydrobia ulvae* in the Ythan estuary.' *J. mar. biol. Ass. UK*, **51,** 423–37.

Engelmann, M. D. (1966). 'Energetics, terrestrial field studies and animal productivity.' In *Advances in Ecological Research*, Vol. 3. Ed. J. B. Cragg. London and New York: Academic Press.

Gorman, M. L. and Milne, H. (1971). 'Seasonal changes in the adrenal steroid tissue of the Common Eider *Somateria mollissima*, and its relation to organic metabolism.' *Ibis*, **113,** 218–28.

Goss-Custard, J. D. (1966). 'The feeding ecology of Redshank, *Tringa totanus* (L.) in winter on the Ythan estuary, Aberdeenshire.' Unpublished Ph.D. Thesis, University of Aberdeen.

Goss-Custard, J. D. (1969). 'The winter feeding ecology of the Redshank, *Tringa totanus*.' *Ibis,* **111,** 338–56.

Goss-Custard, J. D. (1970). 'The responses of Redshank (*Tringa totanus* (L.)) to spatial variations in the density of their prey.' *J. Anim. Ecol.*, **39,** 91–113.

Healey, M. C. (1971). 'The distribution and abundance of sand gobies, *Gobius minutus*, in the Ythan estuary.' *J. Zool., Lond.*, **163,** 177–229.

Heppleston, P. B. (1971). 'Feeding techniques of the Oystercatcher (*Haematopus ostralegus*).' *Bird Study*, **18,** 15–20.

Leach, J. H. (1971). 'Hydrology of the Ythan estuary with reference to distribution of major nutrients and detritus.' *J. mar. biol. Ass. UK*, **51,** 137–57.

McNeill, S. and Lawton, J. H. (1970). 'Annual production and respiration in animal populations.' *Nature, Lond.*, **225,** 472–4.

Marriott, R. W. (1966). 'The food and feeding of the Eider population on the Ythan estuary, Aberdeenshire, in Autumn and Winter.' Unpublished Thesis, University of Aberdeen.

Milne, H. (1963). 'Seasonal distribution and breeding biology of the Eider, *Somateria mollissima* L., in north-east of Scotland.' Ph.D. Thesis, University of Aberdeen.

Milne, H. (1968). 'The Eider-mussel food link in the community of the Ythan estuary.' *Rep. Challenger Society*, **3,** No. XX, 31.

Norton-Griffiths, M. (1967). 'Some ecological aspects of the feeding behaviour of the Oystercatcher, *Haematopus ostralegus*, on the Edible Mussel *Mytilus edulis*.' *Ibis*, **109,** 412–24.

Odum, E. P. (1961). 'The role of tidal marshes in estuarine production.' Contribution No. 29 from the University of Georgia Marine Institute. N.Y. State Conservation Dept, Division of Conservation Education.

Odum, E. P. (1971). *Fundamentals of Ecology*. 3rd ed. Philadelphia: Saunders.

Petrusewicz, K. (Ed.) (1967). *Secondary Productivity of Terrestrial Ecosystems*. Vol. 2. Warsaw: Polish Academy of Sciences.

Walton, K. (1966). 'The Ythan Estuary.' In *Geography as Human Ecology*. Ed. S. R. Eyre and G. R. J. Jones, 30–54. London: Arnold.

Young, C. M. (1964). 'An ecological study of the Common Shelduck (*Tadorna tadorna* L.) with special reference to regulation of the Ythan population.' Ph.D. Thesis, University of Aberdeen.
Young, C. M. (1970). 'Territoriality in the Common Shelduck, *Tadorna tadorna*.' *Ibis*, **112,** 330–5.

APPENDIX

Fauna list of the Ythan estuary

This list is not complete and is heavily biased in favour of some groups. Similarly the status ratings are difficult to interpret precisely and are given as very general indices. The list may be useful in illustrating the variety of species occurring in the Ythan estuary.

Common name	*Scientific name*	*Rare*	*Frequent*	*Common*
AVES				
(i) *Waterfowl*				
Barnacle Goose	*Branta leucopsis*	★		
Greylag Goose	*Anser anser*			★
White-fronted Goose	*Anser albifrons*	★		
Bean Goose	*Anser fabalis*	★		
Pink-footed Goose	*Anser brachyrhynchus*			★
Mute Swan	*Cygnus olor*			★
Whooper Swan	*Cygnus cygnus*		★	
Shelduck	*Tadorna tadorna*			★
Mallard	*Anas platyrhynchos*			★
Teal	*Anas crecca*	★		
Wigeon	*Anas penelope*		★	
Tufted Duck	*Aythya fuligula*		★	
Scaup	*Aythya marila*	★		
Pochard	*Aythya ferina*	★		
Eider	*Somateria m. mollissima*			★
Long-tailed Duck	*Clangula hyemalis*	★		
Goldeneye	*Bucephala clangula*	★		
Red-breasted Merganser	*Mergus serrator*		★	
Goosander	*Mergus merganser*	★		
(ii) *Waders*				
Oystercatcher	*Haematopus ostralegus*			★
Ringed Plover	*Charadrius hiaticula*			★
Golden Plover	*Pluvialis apricaria*		★	
Grey Plover	*Pluvialis squatarola*	★		
Lapwing	*Vanellus vanellus*			★
Turnstone	*Arenaria interpres*			★
Little Stint	*Calidris minuta*	★		
Dunlin	*Calidris alpina*		★	
Curlew Sandpiper	*Calidris ferruginea*	★		

Common name	*Scientific name*	*Rare*	*Frequent*	*Common*
AVES—contd.				
(ii) *Waders*—contd.				
Knot	*Calidris canutus*		★	
Sanderling	*Calidris alba*	★		
Ruff	*Philomachus pugnax*	★		
Redshank	*Tringa totanus*			★
Greenshank	*Tringa nebularia*	★		
Common Sandpiper	*Tringa hypoleucos*	★		
Bar-tailed Godwit	*Limosa lapponica*	★		
Curlew	*Numenius arquata*		★	
Snipe	*Gallinago gallinago*	★		
Jack Snipe	*Lymnocryptes minima*	★		
Spoonbill	*Platalea leucorodia*	★		
Heron	*Ardea cinerea*		★	
(iii) *Gulls*				
Black-headed Gull	*Larus ridibundus*			★
Herring Gull	*Larus argentatus*			★
Great Black-backed Gull	*Larus marinus*		★	
Common Gull	*Larus canus*			★
Kittiwake	*Rissa tridactyla*		★	
Sandwich Tern	*Sterna sandvicensis*			★
Common Tern	*Sterna hirundo*			★
Arctic Tern	*Sterna paradisea*			★
Roseate Tern	*Sterna dougallii*	★		
Little Tern	*Sterna albifrons*		★	
Arctic Skua	*Stercorarius parasiticus*	★		
Great Skua	*Stercorarius skua*	★		
(iv) *Others*				
Cormorant	*Phalacrocorax carbo*			★
Little Grebe	*Podiceps ruficollis*	★		
Black-necked Grebe	*Podiceps nigricollis*	★		
PISCES				
Common Eel	*Anguilla anguilla*			★
Greater Pipe Fish	*Syngnathus acus*	★		
Lesser Sand Eel	*Ammodytes tobianus*		★	
Viviparous Blenny	*Zoarces viviparus*		★	
Common Blenny	*Blennius pholis*		★	
Butterfish	*Centronotus gunnellus*			★
Stickleback, 15-Spined	*Gasterosteus spinachia*	★		
Stickleback, 3-Spined	*Gasterosteus aculeatus*		★	
Common Goby	*Gobius minutus*			★
Goby	*Pomatoschistus microps*		★	
Flounder	*Platichthys flesus*			★

Common name	Scientific name	Rare	Frequent	Common
PISCES—*contd.*				
Plaice	*Pleuronectes platessa*	★		
Bullhead	*Agonus cataphractus*	★		
Sea Scorpion	*Cottus scorpius*		★	
Lumpsucker	*Cyclopterus lumpus*	★		
Sea Trout	*Salmo trutta*			★
Sprat	*Clupea sprattus*		★	
Herring fry	*Clupea harengus*	★		
Mullet	*Mugil chelo*	★		
Ray's Bream	*Brama rayi*	★		
Saithe	*Pollachius virens*		★	
Bass	*Dicentrarchus labrax*	★		
Pollock	*Pollachius pollachius*	★		
INVERTEBRATA				
(i) *Mollusca*				
Mussel	*Mytilus edulis*			★
Horse mussel	*Modiolus modiolus*	★		
Cockle	*Cerastoderma edule*		★	
	Cerastoderma lamarcki	★		
	Macoma balthica			★
	Scrobicularia plana	★		
	Mya arenaria		★	
Periwinkle	*Littorina littorea*			★
	Littorina littoralis		★	
	Littorina rudis		★	
	Hydrobia ulvae			★
	Alderia modesta	★		
(ii) *Crustacea*				
	Balanus balanoides		★	
	Ligia oceanica		★	
	Gammarus salinus			★
	Gammarus zaddachi			★
	Gammarus pulex		★	
	Gammarus locusta	★		
	Gammarus duebeni			★
	Bathyporeia pilosa		★	
	Marinogammarus marinus			★
Sandhopper	*Talitrus saltator*			★
	Orchestia gammarella			★
	Talorchestia deshaysii		★	
	Corophium volutator			★
	Hyale nilssoni			★
Opossum shrimp	*Praunus flexuosus*		★	
	Neomysis integer			★
Shore Crab	*Carcinus maenas*			★

Common name	Scientific name	Rare	Frequent	Common
INVERTEBRATA—*contd.*				
(ii) *Crustacea*—contd.				
Common Shrimp	*Crangon vulgaris*			★
	Cyclops strenuus		★	
	Cyclops vulgaris		★	
	Cyclops bisetosus		★	
	Eurytemora affinis		★	
	Eurytemora velox	★		
(iii) *Annelida*				
	Nereis virens		★	
	Nereis diversicolor			★
	Nereis cultrifera	★		
Lugworm	*Arenicola marina*			★
	Monopylephorus irrorata	★		
	Nerilla antennata	★		
	Tubifex newaensis		★	
	Tubifex pseudogaster	★		
	Tubifex costatus	★		
	Enchytreus sp.		★	
	Lumbriculus variegatus	★		
	Paranais littoralis		★	
	Dero sp.	★		
	Nais elinguis	★		
	Peloscolex benedeni		★	
(iv) *Cnidaria*				
Common jellyfish	*Aurelia aurita*	★		
Sea gooseberry	*Pleurobrachia pileus*		★	
	Cyanea capillata	★		
(v) *Protozoa*				
	Elphidium (*Polystomella*) sp.			★
	Rotalia sp.			★
(vi) *Others*				
	Lineus ruber	★		
	Lineus gesserensis			★
	Priapulus caudatus	★		

7

The Importance of Estuaries to Commercial Fisheries

P. R. WALNE

INTRODUCTION

A consideration of the role of estuaries in the fisheries of the United Kingdom shows that, with one exception, the importance is mainly for shellfish (molluscs and crustacea) rather than for fish. The exception is the migratory fish—salmon, sea trout and eels—which have to pass from lakes and rivers, through estuaries, to the sea in order to complete their life cycle. Although the estuary plays only a small part in the life of these species it is an essential link which must not be broken. The value of these fisheries is considerable; for example, the first-sale value of salmon landed in the United Kingdom in 1970 was about £2,256,000. The well-known recreational value of this species makes it economically important to the community in some areas. This may be compared with the total value of £500,000 for the predominantly estuarine fisheries of cockles, mussels, oysters and winkles.

The estuaries of the United Kingdom are small compared with some in other countries; for example, we have nothing to compare with Chesapeake Bay on the north-east coast of the United States, which has a length of one hundred miles and a width of twenty-five miles at the mouth, and is sufficiently extensive to have large areas of stable, but reduced, salinity. In such areas actively moving species of fish and crustacea, as well as the more sedentary forms, can spend their entire life in an area of reduced but stable salinity. The value of the oyster crop alone was $13,600,000 in 1962.

FACTORS AFFECTING THE PRODUCTION OF SHELLFISH

Some of the shellfish of the United Kingdom are sufficiently localised, and the collection of statistics sufficiently reliable, for estimates to be made of the fishery on a unit area basis (Table 1). The areas which have been used as a basis for calculation are the legally defined limits of the fishery; in each case the fisheries are the subject of a Several or Regulating Order made under the authority of the Sea Fishery Acts (here 'Several' has its older, legal meaning of private and individual). The boundaries will have been drawn in relation to the fishery and also with regard to suitable natural marks to define the required grounds. Consequently the legally defined fishery will

TABLE 1

ANNUAL YIELD PER HECTARE OF SHELLFISH FROM SELECTED FISHERIES IN THE UNITED KINGDOM. ALL FIGURES ARE YEARLY AVERAGES FOR THE PERIODS SHOWN, EXCEPT FOR THOSE FOR THE MENAI STRAITS, WHICH ARE THE DATA FOR ONE YEAR ONLY

Species	*Period*	*Area* (*ha*)	*Yield* (*kg/ha*)
COCKLES			
Burry Inlet	1966–70	1 030	3 780
Boston	1968–70	7 776	562
MUSSELS			
North Norfolk	1968–70	65·6	7 826
Menai Straits	1969–70	780	2 235
Conway	1966–70	777	556
Boston	1968–70	7 776	134
OYSTERS			
Fal	1966–70	706	319

include some areas which are quite unsuitable for the species concerned, while due to natural fluctuation in recruitment to the population, not all suitable areas are fully productive at all times.

Despite these difficulties the figures presented in Table 1 suggest that a reasonably productive estuary can yield about 1 000 kg (live weight) of either cockles or mussels per hectare. The figure for oysters from the River Fal is low because of failures in annual recruitment, but since, on a weight for weight basis, oysters are worth over thirty times as much as mussels, the economic yield per hectare is in favour of the Fal oyster fishery. The boundaries of the Boston fishery in the Wash are very wide and include both mussel and cockle grounds, and this accounts for the relatively poor yield. The Menai Straits mussel fishery includes a cultivated area and this enhances the average yield. The north Norfolk mussel fishery is completely cultivated and the landings illustrate the potential of such small cultivated areas, in this case nearly 8 000 kg/ha/year.

As the areas of a fishery become more sharply defined, so the size of the standing stock and the crop rises, because less unproductive ground is included in the calculation, until the point is reached where full cultivation is practised and there are no unproductive areas.

TABLE 2

THE DRY WEIGHT OF MEAT IN THE STANDING STOCK OF VARIOUS COMMERCIALLY IMPORTANT BIVAVLES IN NORTHERN EUROPE

Species	*Dry meat* (g/m^2)	*Area at this density (ha)*	*Locality*	*Source*
Cockles (*Cardium edule*)	120	515	Burry Inlet	A. Franklin (pers. comm.)
Mussels (*Mytilus edulis*)	1 940	9	Menai Straits	P. J. Dare (pers. comm.)
Mussels (*Mytilus edulis*)	600	40	Morecambe Bay	P. J. Dare (pers. comm.)
Mussels (*Mytilus edulis*)	560	12	Danish Wadden Sea	Theisen (1968)
Mussels (*Mytilus edulis*)	210	32	Wexford	Meaney (1970)
Oysters (*Ostrea edulis*)	200	–		P. R. Walne (unpublished data)
Quahog (*Mercenaria mercenaria*)	206	3	Southampton Water	P. R. Walne (unpublished data)

Examples of the size of the standing stock in natural populations of cockles and mussels is given in Table 2. Clearly, crops of 100 g or more of dry flesh per m^2 can be expected. Mussels, because of their ability to raise themselves above the bottom, can reach substantially higher densities over extensive areas.

Theisen (1968) studied the growth and survival of large-scale transplants (145–1 926 tons per plot) of mussels in the Danish Wadden Sea. The mussels were laid at densities of 4–8 kg/m^2, which corresponds to about 280–560 g dry meat/m^2. Although the individual mussels had doubled in weight in the six to twelve months before harvesting, the mortality rate was such that the stock per unit area had not increased.

CONDITIONS FOR SHELLFISH CULTIVATION

It will have been noted that in the previous paragraph reference has been made to cultivation. This is the most important aspect of the interrelation of shellfisheries and estuaries, and it is brought about not only by the salinity regime of estuaries but also by the shelter which they provide. In this context examples can be quoted from one extreme, where shelter plays a dominant part and reductions in salinity play only an irregular role in modifying the environment, to the other extreme, where the salinity pattern is at least as important as shelter, if not more so. For instance, the estuaries of the Blackwater, Crouch and Roach in Essex formerly supported substantial oyster industries. In these cases the inflow of fresh water can be so low in the summer months (Talbot, 1967) that it is insufficient to balance evaporation and the salinity can rise to a higher level than that occurring in the sea outside. This is because these estuaries are former outlets of the River Thames and they are now much larger than their drainage basin warrants (Sheldon, 1968). At the other extreme the River Conway can be quoted as an example of a river rising in mountainous country and where even at the mouth the salinity is substantially reduced at every low tide. Poole Harbour, a most productive area, has been formed by a drowned low-lying valley, the mouth of which has become partially barred by sand-dune formations arising on either side of the mouth.

Cockles, mussels and other commercially important bivalves release eggs and sperm into the surrounding water; after fertilisation, these develop into free-swimming larvae. Oysters release partially developed larvae into the water; after a period of a week or more these settle to the bottom, where they undergo a metamorphosis into the adult form. (Most species of oysters then take two to five years to reach a commercial size.) During that week, although the larvae are swimming they can do little more than float passively in the current, and the opportunity for dispersion is considerable. In this case the physical form of an estuary can play a part in reducing dispersion. For example, oyster larvae released by breeding stock in the upper part of the River Crouch will not reach the mouth—a distance of fourteen miles—on the ebb tide (Knight-Jones, 1952). Similarly, I have found that concentrations of oyster larvae in the upper part of Poole Harbour do not reach the mouth and pass out into Poole Bay during the tidal cycle. It has been suggested (Bousfield, 1955; Wood and Hargis, in press) that in some estuaries planktonic larvae of various species can take advantage of the inward-flowing, underlying, more saline water and so move inland and away from the mouth. Some estuaries are apparently too short. The River Yealm in South Devon has been stocked with oysters for many years and they grow and fatten very well (Walne, 1970), but a spatfall has never been recorded there.

There are numerous examples of bivalve larvae settling in parts of estuaries which are unsuitable for their full development to adults, and it is on such characteristics that the simpler cultural methods depend. For example, oyster larvae formerly settled in large numbers on stones and shells at the heads of the Essex estuaries. This may be partly due to the concentrating mechanisms referred to in the previous paragraph and partly to the abundance of suitable clean surfaces for attachment in these areas. However, these grounds are subjected to sharply reduced salinities and lowered temperatures in the winter months, and these conditions are inimical to oyster survival. This is also probably the reason why the stones and shells are free from other competing organisms. Likewise, mussels settle at levels where the tidal exposure is too great for fast growth or in places where they can be dislodged by the winter storms. The cultivated oyster fisheries of France and America provide other examples of the necessity of moving young individuals to more favourable areas.

A further source of juveniles for stocking purposes is provided by the development of hatcheries. Techniques are now available whereby large numbers of the larvae of various species of oysters and clams can be reared under controlled conditions. Filtered and partially-sterilised sea water is used, and cultures of selected species of algae are added for food. Suitable surfaces for settlement have to be provided for oyster larvae when they reach the time of metamorphosis; plastic sheets with a matt surface are generally used for this. Immediately after metamorphosis the spat are removed from the plastic and transferred to trays of running sea water until they reach a few millimetres in size. They are then transferred to trays suspended in the sea until they are large enough to be planted on the sea bed. The trays can either be on posts near low water of spring tides or they can be suspended from a raft; in either case a sheltered situation is essential.

The cultivation of bivalves in the sea is made possible by the animals' inability to move more than a very short distance. Unless moved by currents they remain where they are placed and can be recovered by the person who put them there in the first place. From this it follows that an area of the sea bed may be farmed, the stock of animals being tended by individuals or commercial concerns. This is legally recognised by the granting of private rights to areas of the sea bed under the Several Order procedure operated under the authority of the Sea Fishery Acts.

Cultivation on the sea bed requires that the bottom soil should be suited to the species concerned. This factor is particularly critical for the oyster, since it has no power of movement and if it is moved, due to instability of the bottom soil, or if it is covered by a deposition of silt, the oyster is unable to recover a favourable position. An extreme example is provided by the oyster fishery of the Oosterschelde which formerly produced 30 000 000 oysters annually. An important contributing factor to this production was the existence of extensive areas of firm peaty bottom soil resulting from the recent submergence of the land. Mussels may be laid on a variety of firm substrates, but difficulties arise if they are exposed to strong winds or currents. Mussels can grow at a high density (Table 2) and attach to each other and to the substrate by means of their byssus threads. In time a continuous mat of mussels is formed, amongst which their faeces and pseudo-faeces accumulate. To avoid being smothered the mussels detach from the substrate and move upwards so as to keep on top of the mud. My colleague, Dr Dare, has made some observations on this phenomenon in Morecambe Bay, where he found that in 1968 a mussel carpet moved upwards a distance of 50 cm between May and September; it was less in 1969 and 1970, but in 1971 the movement again reached 50 cm. Such a situation is very unstable because by late summer the mussels are resting on a thick layer of soft mud, and in an exposed situation, such as in Morecambe Bay, they will be washed away by the winter gales. Theisen (1968) describes a similar situation in the Danish Wadden Sea.

The use of rafts allows exploitation of the whole water column and keeps the stock away from any influence of the bottom. Culture of this type is practised to only a very minor degree in England; the main centres of raft culture are in north-west Spain for mussels and in Japan for oysters. The methods employed require sheltered conditions and entail considerable labour, but the yield can be very high.

The Spanish mussel culture is centred in the sunken valleys of Galicia. These valleys, 10–15 miles long and 3–6 miles wide, are sufficiently deep and sheltered for numerous rafts to be moored in water which is about 10 m deep; although in this case salinity reduction is minimal, they illustrate a culture method which requires sheltered conditions. The rafts support 500–1 000 vertical ropes, each 10 m long. The mussels settle naturally on the ropes and, after various cultural procedures, are ready for sale when 12–18 months old. An average raft can produce 40–60 metric tonnes per year (30 tons of drained meat). This industry started about twenty-five years ago and has steadily increased until there are now over 3 000 rafts. The yield from them has lifted Spain to the position of the leading country in the world for mussel production, and they form an outstanding illustration of the rate at which a cultivated fishery may develop.

In regions where the tidal range is small, large shallow lagoons may be formed at the mouths of streams and rivers. The fresh water inflow may be enhanced by submerged springs, and the water may be sufficiently rich in nutrients to significantly increase the productivity of the area (Korringa and Postma, 1957). Shallow lagoons of this type may be sufficiently sheltered for the culture of mussels on vertical ropes hanging from horizontal ropes or on bars suspended from stakes driven into the bottom. Lake Varano on the east coast of Italy produces annually 20 000 tons of mussels by this method; in addition, substantial quantities of eels and other fish are caught.

Japan produces about 250 000 metric tonnes of whole oysters (35 000 tons of oyster meat, nearly all *Crassostrea gigas*) per year, and they are nearly all grown in suspended culture. Clean shells are threaded on wires and hung in the water to act as settlement surfaces for naturally occurring oyster spat. Regions of reduced salinity give the best results. After settlement the number of spat on each shell is reduced to four or five, and the shells are re-threaded with spacers between to allow plenty of room for circulation of the water. The wires may be suspended from rafts but many are suspended from lines stretched between moored floats. Growth is rapid and many oysters are ready for sale at six to twelve months of age.

The previous paragraphs have reviewed some of the methods which are used for the exploitation of shellfish. What are the important factors which influence these fisheries?

Primary production

There are not many estimates of the primary (photosynthetic) production in estuaries. Those which have been made suggest that they are probably in the range corresponding to the values of 100 g of carbon fixed/m^2/year in shallow coastal water and 300 g of carbon/m^2/year in areas of coastal upwelling. The shallow depth and strong currents ensure that the water column is well mixed and that nutrients are continuously transported to the surface layers where photosynthesis is taking place. The mixing process prevents the formation of the thermocline which, in deeper water, inhibits the circulation of nutrients in the summer months. In some estuaries the fresh water flowing in may be an important source of fresh nutrients. On the other hand, the turbulence pattern and the frequent occurrence of large intertidal flats of soft deposits lead to a substantial load of particulate matter in the water; the data reviewed in Table 3 suggest that values of 10 mg/l, and more, frequently occur in areas important to shellfisheries. This will reduce the penetration of light and the depth of the photic zone. The photosynthetic flora attached to the bottom soil in intertidal or shallow water zones may make a considerable contribution to production. Grøntved (1960) calculated that the microbenthos was responsible for fixing 116 g carbon/m^2/year in shallow lagoons in Denmark; this was about four times greater than production by the phytoplankton in the overlying water. Similarly, it has been suggested for many years that

TABLE 3

THE SUSPENDED MATTER IN THE WATER OVER SOME SHELLFISH GROUNDS

Area	*Dry matter (mg/l)*	*Authority*
OYSTER GROUNDS		
River Blackwater, England	10 –100	Waugh (1966)
River Belon, France	6 – 18	Marin (1971)
MUSSEL GROUNDS		
Menai Straits, Wales	0·5– 50	Buchan *et al.* (1967)
COCKLE AND MUSSEL GROUNDS		
Wadden Sea, Netherlands	7 – 60	Postma (1954)
Wadden Sea, Denmark	11 –278	Gry (1942)

the decomposition of macrophytes is an important nutrient source. In turbulent conditions both these sources of food would be available to suspension feeders.

On an ash-free basis the carbon content of marine organisms is usually in the range of 45–55 per cent of the dry meat weight (Strickland, 1965), and therefore the production corresponds to about 600 g of organic matter/m^2/year. Bivalves consume unicellular algae with an assimilation efficiency which is unknown but which is likely to be similar to that found in other marine animals. Therefore, of the food that they eat probably 10–15 per cent is converted into flesh. If all the annual production of the rich estuary postulated above were converted into commercially important bivalves, this would allow a production of 60–90 g/m^2 of flesh per year. This will not occur because of the large number of other animals, both in the plankton and on the bottom, which will be competing with the bivalves for the food. In practice only a small proportion of the 600 g/m^2 of organic matter produced per year will be available to the shellfish. Table 2 gives some details of the standing stock of bivalves in some areas. The cockle population from the Burry Inlet was two years old, giving an annual production of 60 g/m^2 organic matter. The mussels in Morecambe Bay were less than a year old at the time of measurement, giving an annual production of not less than 900 g/m^2/year. The mussels in Denmark had doubled their weight and therefore production had been at least 200 g/m^2/year in this example. Ryther (1969) calculates that the annual production of a Spanish mussel raft is no less than 10 kg of dry meat/m^2 on an areal basis; Japanese oyster production is about one-tenth of this.

These calculations emphasise that the productivity of many shellfisheries depends on the primary production of areas which are much more extensive than those occupied by the fishery. The detailed work of Postma (1954) in the Netherlands Wadden Sea has shown that about half the fixed carbon originates within the Wadden Sea, and about half comes in from the North Sea. It is brought in by the tide and trapped by the abundance of filter-feeding organisms. Not only do they use it for their own growth and metabolism, but much of it is thrown to the bottom in the more stable form of faeces and pseudo-faeces, where it is continually reworked by benthic animals. An example of the rapidity of this type of accumulation has been quoted from observations made in Morecambe Bay.

Salinity changes

A fluctuating salinity is the most obvious source of stress in an estuary, and regions of high stress are characterised by a reduced number of species which may, because of diminished competition, reach very large numbers. Some species of bivalves which are commercially important in the United Kingdom, for example the oyster *Ostrea edulis*, the mussel *Mytilus edulis*, and the cockle *Cardium edule*, are able to withstand to varying degrees the stresses imposed in an estuary. None of these species has a requirement for reduced salinity, but it is in these conditions that reduced competition and lowered predation are most frequently encountered. A very clear example of an application of this principle is given by the Dutch mussel fishery in the Wadden Sea, where it is the practice to locate beds of mussel spat while they are still small, and hence vulnerable to predation, and transfer them to areas of reduced salinity. The mussels grow well but predation is greatly reduced because the salinity cannot be tolerated by the starfish *Asterias* and the shore crab *Carcinus*. Heavy spatfalls of mussels occur from time to time in full salinity coastal waters of the United Kingdom, but they are usually quickly destroyed by flatfish, starfish and crabs. An example has been given by Meaney (1970) who estimated that one bed off Pass Head, County Wexford, contained about 1 000 tons of mussels at an average density of 3 kg/m^2. Such spatfalls only survive on structures raised above the bottom where they are free from predation; for example, navigational buoys and light-vessels are often densely encrusted with mussels.

Currents

The rate of water movement can vary widely over quite small areas and this affects the degree of scour, the shifting of the bottom soil and the transport of food to sessile organisms.

If the velocity of the water is too great, the physical bombardment of sessile organisms by sand particles will cause frequent contraction of the feeding organs and lead to reduced growth. This is a factor of importance at the time of settlement, since the free-swimming larva has to be able to attach to the chosen surface long enough for metamorphosis to proceed. An additional hazard is provided by the scouring action of sand particles which, by their size and momentum, break the comparatively weak larval shell and scour away the contents (Shelbourne, 1957). The section on production has shown how many populations are dependent on the transport of the products of primary production to their grounds.

Temperatures

The shallow waters of bays and creeks have a greater annual temperature amplitude than that of the more stable environment of the sea. The low temperatures of winter restrict the cultivation of some species; *Ostrea edulis* will not regularly survive intertidally or in very shallow water on the east coast of England. The effect of an exceptional winter is well documented (Crisp, 1964). On the other hand, the rapid warming in summer may be sufficient to trigger the spawning of species which would not otherwise do so. It is characteristic that spawning may require a higher temperature than is necessary for the development of the larvae, metamorphosis and growth to the adult. The establishment of a population of *Crassostrea gigas* in Pendrell Sound in British Columbia (Quayle, 1969) is a clear example of this phenomenon on a large scale. An illustration on a smaller scale is given by imported stocks of *Crassostrea angulata* in some English estuaries, and the establishment of the North American clam *Mercenaria mercenaria* in Southampton Water is believed to be due to this phenomenon, reinforced in this case by the warm-water effluent of a nearby power station (Ansell, 1963).

The activities of man

Among the hazards faced by estuarine fisheries are those caused by man. The sheltered nature of estuaries makes them suitable for the development of ports and so they tend to become centres of communication. Moreover, the top of the estuary is often the first practicable crossing place on a long indentation of the coastline. For these reasons many estuaries are now centres of large populations. In these circumstances pollution will be a problem, and this may be so dominant—especially if it is of industrial origin—as to deprive the estuary of all fishery interest. Pollution by domestic wastes in more rural areas may make it necessary for shellfish to undergo a purification process before they are fit for sale as food. In some circumstances pollution of this type may cause advanced eutrophication. The occurrence of this hazard over the clam fisheries in Great South Bay, on Long Island, which was due to the drainage from duck farms, has been well documented (Ryther and Dunston, 1971).

Other activities of man are less clear in their effect. The demands of navigation lead to the dredging of channels, which alters the current pattern and may increase the silt load in the water. Expanding populations lead to demands for the reclamation of wet lands. The potential importance of these as a source of the primary production of organic matter in estuaries has been emphasised earlier but it is a difficult problem to quantify. It has also been suggested that increased abstraction of water from the river system feeding the estuary may be sufficient to alter the salinity pattern and hence the ecology of the area.

Recreational activities compete with shellfisheries in a manner analogous to that occurring on farmland. Intertidal shellfish beds cannot be walked upon without harm, and a large boat which accidentally goes aground can press several square meters of shellfish into the soil. Moorings are a hazard to cultivation procedures, apart from interfering with areas of the sea bed. Shellfish rafts for suspended culture need deep water, but such channels are often required for commercial traffic, while trays of shellfish, whether held above the bottom on posts or in rafts, may be a hazard to small boats.

REFERENCES

Ansell, A. D. (1963). '*Venus mercenaria* (L.) in Southampton Water.' *Ecology*, **44,** 396–7.

Bousfield, E. L. (1955). 'Ecological control of the occurrence of barnacles on the Miramichi Estuary.' *Bull. natn. Mus. Can.*, No. 137, 69 pp.

Buchan, S., Floodgate, G. D. and Crisp, D. J. (1967). 'Studies on the seasonal variation of the suspended matter in the Menai Straits. 1. The inorganic fraction.' *Limnol. Oceanogr.*, **12,** 419–31.

Crisp, D. J. (Ed.) (1964). 'The effects of the severe winter of 1962–63 on marine life in Britain.' *J. Anim. Ecol.*, **33,** 165–210.

Grøntved, J. (1960). 'On the productivity of microbenthos and phytoplankton in some Danish fjords.' *Meddr Danm. Fisk.-og Havunders.*, N.S., **3,** 55–92, 3 plates.

Gry, H. (1942). 'Das Wattenmeer bei Skallingen. Physiographisch-biologische Untersuchung eines dänischen Tidengebietes. No. 1. Quantitative Untersuchungen über den Sinkstofftransport durch Gezeitenströmungen.' *Folia geogr. dan.*, **2,** No. 1, 138 pp.

Knight-Jones, E. W. (1952). 'Reproduction of oysters in the Rivers Crouch and Roach, Essex, during 1947, 1948 and 1949.' *Fishery Invest., Lond.*, Ser. 2, **18,** No. 2, 48 pp.

Korringa, P. and Postma, H. (1957). 'Investigations into the fertility of the Gulf of Naples and adjacent salt water lakes, with special reference to shellfish cultivation.' *Pubbl. Staz. zool. Napoli*, **29,** 229–84.

Marin, J. (1971). 'Étude physico-chimique de l'estuaire du Belon.' *Rev. Trav. Inst. Pêch. marit.*, **35,** 109–56.

Meaney, R. A. (1970). 'Investigations on a seed mussel bed off the coast of Wexford during 1970.' *Irish Sea Fisheries Board, Resource Record Paper*, 11 pp. (mimeo).

Postma, H. (1954). 'Hydrography of the Dutch Wadden Sea. A study of the relations between water movement, the transport of suspended materials and the production of organic matter.' *Archs néerl. Zool.*, **10,** 405–511.

Quayle, D. B. (1969). 'Pacific oyster culture in British Columbia.' *Bull. Fish. Res. Bd Can.*, No. 169, 192 pp.

Ryther, J. H. (1969). 'The potential of the estuary for shellfish production.' *Proc. natn. Shellfish Ass.*, **59,** 18–22.

Ryther, J. H. and Dunston, W. M. (1971). 'Nitrogen, phosphorus and eutrophication in the coastal marine environment.' *Science N.Y.*, **171,** 1008–13.

Shelbourne, J. E. (1957). 'The 1951 oyster stock in the Rivers Crouch and Roach, Essex.' *Fishery Invest., Lond.*, Ser. 2, **21,** No. 2, 27 pp.

Sheldon, R. W. (1968). 'Sedimentation in the estuary of the River Crouch, Essex, England.' *Limnol. Oceanogr.*, **13,** 72–83.

Strickland, J. D. H. (1965). 'Production of organic matter in the primary stages of the marine food chain.' In *Chemical Oceanography*, Vol. 1 (Ed. J. P. Riley and G. Skirrow), pp. 477–610. London and New York: Academic Press. 712 pp.

Talbot, J. W. (1967). 'The hydrography of the estuary of the River Blackwater.' *Fishery Invest., Lond.*, Ser. 2, **25,** No. 6, 92 pp.

Theisen, B. F. (1968). 'Growth and mortality of culture mussels in the Danish Wadden Sea.' *Meddr Danm. Fisk.-og Havunders.*, N.S., **6,** 47–78.

Walne, P. R. (1970). 'The seasonal variation of meat and glycogen content of seven populations of oysters *Ostrea edulis* L. and a review of the literature.' *Fishery Invest., Lond.*, Ser. 2, **26,** No. 3, 35 pp.

Waugh, G. D. (1966). 'Turbidity estimation using the Secchi disc.' MAFF, Bradwell Investigations, Report No. 3, 11 pp. (mimeo.).

Wood, L. and Hargis, W. J. 'Factors associated with the transport and retention of bivalve larvae in a tidal estuary.' *Fourth European Symposium on Marine Biology*, 1969 (in press).

8

The Effects of Pollution on the Thames Estuary*

M. J. Barrett

INTRODUCTION

Studies of water quality in the Thames Estuary have been carried out by the Water Pollution Research Laboratory not only to document the conditions to be found in that body of water but also to develop a method capable of predicting the results of any combination of conditions likely to be encountered in the future. To this end, several lines of study were adopted, including (*a*) regular determinations of water quality throughout the estuary over a number of years; (*b*) statistical analysis of a long series of observations of past conditions in the estuary, kindly made available by the London County Council, now the Greater London Council; (*c*) study of the sources of polluting substances; (*d*) determination of the rates at which oxygen was removed from the water and replenished from the atmosphere and from other sources; and (*e*) the development of a mathematical model to represent the movement and dispersion of substances discharged into the estuary.

The methods of calculation developed by this study were checked by comparing the model's predictions with the observed distribution of dissolved oxygen and both ammoniacal and oxidised nitrogen for each three-month period from 1950 to 1961. In this way it was shown that the methods were sufficiently accurate to be usefully applied to calculate the effects of changing variables, such as temperature and freshwater flow, and to forecast the effects of changes in the location

* *Editors' Note*—Owing to unforeseen circumstances, it has proved impossible to include in this volume a full written version of the paper given by Mr Barrett at the symposium. What follows is a detailed summary of his paper prepared by the editors from a verbatim transcript made and kindly put at their disposal by Dr N. J. Holmes. This summary has been checked and approved by Mr Barrett.

and strength of polluting discharges. The recent work which has been carried out still continues to support the accuracy of the methods developed.

This paper examines the long term changes which have taken place in the distribution of dissolved oxygen in the Thames over the last 80 years, especially with regard to the recent major improvements in the treatment of London's sewage. Other aspects of this study, including the methods of prediction, have been dealt with elsewhere.

Before examining these changes, however, two points need to be mentioned. First, adjustments are necessary to allow for the state of the tide, since the oxygen content of samples taken only a few hours apart at the same point in an estuary can be very different from one another. This is not, in general, because of the rate of change of the oxygen content of the water, but because the samples have been taken from different bodies of water. Accordingly, if misleading results are to be avoided, it is necessary to use only those data which relate to the same tidal state or to make allowances for the movement of water during a tidal cycle. Secondly, one can not simply consider the oxygen content at a particular point and say, for example, that the oxygen is lower in the summer because the temperature is high, or that because the oxygen content is less than during the previous year the condition of the water is deteriorating, as this may, in fact, be a measure of the effect of different flow rates. Therefore, flow rates must also be taken into consideration. The adjustments which have been developed for these phenomena in the Thames also appear to apply quite well to other estuaries which, like the Thames, are only slightly stratified.

THE DISTRIBUTION OF DISSOLVED OXYGEN, 1893–1970

The Greater London Council never samples the whole estuary in a single day and it is necessary to take average values over a period to produce complete oxygen profiles. From various considerations, it was decided for the purposes of this study to obtain average oxygen profiles for each quarter of the year and this has been done for the whole period from 1920 to 1970, as well as for a number of earlier years. In making a quantitative study of the changes that have taken place in the oxygen content both graphical and statistical methods have been used. The period from 1920 onwards was divided into groups of ten years and from the data for average dissolved oxygen at half tide states in a particular decade, the oxygen content was calculated for intervals throughout the estuary. The figures for a particular decade, quarter and position were then plotted against the flow at Teddington.

From the family of curves obtained it is possible to construct sag-curves for particular decades, quarters and flows and by this means to examine the changes which have taken place from one period to another. The following account considers the July to September quarter when the flow at Teddington was about 13 m^3/s (*i.e.* roughly 250 000 000 gallons/day). The earliest curve which it was possible to construct was for 1893 (from the results of an intensive series of surveys made from July 1893 to March 1894 by Dibden). The curves for 1900 and 1905 show little change from this condition. But the curves for 1920–29, 1940–49 and 1950–59 show a progressive deterioration. Although the records show that as early as 1920 dissolved oxygen was absent from particular points for some parts of the year, it was not until 1947 that there was a reach devoid of dissolved oxygen throughout the July–September quarter. The important point about the 1950s is, of course, that sulphide was present in the water. After each winter it appeared first in the centre of the estuary, between 40 km and about 55 km downstream of Teddington Weir, in water of comparatively low salinity when the temperature was about 15°C. At the time that sulphide first appeared there was already in the estuary a reach from 19 km to 25 km long in which there was no dissolved oxygen or nitrate, and once sulphide had been formed it persisted as long as any reach was permanently anaerobic. For instance, in one year sulphide was present at the end of December and it was not until the middle of January, when an increase in flow coincided with a gale, that it could no longer be detected. Following the commissioning in 1960–62 of additional treatment plant at the Beckton sewage works, which is some 48 km below Teddington, conditions began to improve. The final curve, for 1970, shows quite a marked improvement. This is due mainly to a reconstruction of the Crossness works, 52 km below Teddington, and to the improvements made at the Mogden works which discharge a few kilometres below Teddington.

However, at first sight the changes in the past 50 years do not look very great. This is because when the oxygen content falls to low values, the rôle of nitrogen compounds changes. Thus it was concluded after a laboratory study that when the dissolved oxygen falls from about 10% of saturation to zero, the oxidation of ammonia in the water lessens and eventually ceases. Over the same range oxidised nitrogen is reduced, mainly to nitrogen gas. Consequently, other things being equal, a much greater polluting load is required to lower the sag-curve minimum from 10% to 5% of saturation than from 15% to 10% and an even greater increase is needed to reduce it from 5% to zero. In addition, sulphate is reduced under anaerobic conditions and this is a further source of oxygen, albeit a rather undesirable one. All this tends to stabilise the central part of the sag-curve over a wide range of loads.

Twenty years ago, the major concern was the elimination of anaerobic conditions and it is of special interest to examine how the length of the anaerobic reach has changed in recent years. The approximate length of this reach was found for each third day during 1945–65, by plotting out individual oxygen data. Regression of this length on the flow at Teddington was then carried out for each year separately, with the exclusion of flows exceeding about 80 m^3/s. There was no need to treat the 1966–70 data in the same way as anaerobic conditions did not occur then. Conditions remained fairly steady from 1945 to 1948, but there was a marked change from 1949 to 1950 which is attributable to the widespread introduction of packaged synthetic ionic detergents around that time and the consequent lowering of reaeration in the estuary. In 1960, when the new activated-sludge plant came into operation at the Beckton works, there was a marked shortening of the anaerobic reach, and when the new works at Crossness were commissioned in 1964 the anaerobic reach disappeared. This satisfactory condition has continued ever since.

A sensitive measure of the changes which have occurred in the estuary is obtained by considering the seaward end of the sag-curve, for although the reduction in polluting load may not result in much change in the sag-curve minimum, it should in any event produce a shortening of the reach of low oxygen content. At a point 70 km downstream of Teddington, where the effect of fresh-water flow is small or virtually insignificant, the oxygen content of the water fell for about 30 years at a rate of about 1% per year. When the first stage of improvements at Beckton (a new sedimentation plant) was commissioned in 1955 the deterioration seems to have been halted and on the commissioning of a new activated-sludge plant there in 1960 there was an immediate improvement. Since the commissioning of the similar activated-sludge plant at Crossness in 1964 there has been a continuous steady improvement with the result that conditions in the estuary are now back to about what they were in the 1920s.

9

Summarising Review

DON R. ARTHUR

As a zoologist I have always recognised that estuaries are the most dynamic of environments in that the conditions within them vary more widely, rapidly and frequently than elsewhere. But the papers on the chemistry of estuaries and on sedimentation in estuaries given by Mr Phillips and Dr Dyer reveal an apparently ever-increasing complexity. They also contribute to our understanding of some existing biological problems. Mr Phillips showed that an estuary is a region of mixing between two aqueous solutions of very different chemical composition and accordingly it would be incorrect to assume that estuarine waters are simply dilutions of sea water. In this respect he showed, for example, that silicon is less abundant in coastal sea water than in river water and its concentration in estuaries often shows a negative correlation with salinity. When estuaries drain extensive areas of agricultural land, as in the River Ythan described by Drs Milne and Dunnet, the main source of soluble nitrates is land based and the marked seasonality of the occurrence of these compounds reflects agricultural practices in the area. Phosphates, on the other hand, appear to be derived from the inflow of North Sea waters and the levels of phosphate parallel those found in the sea. In estuaries associated with urban areas these salts may also be derived in some quantity from sewage works and a large proportion of them probably have their origin as carriers in domestic products. The point was also made by Mr Phillips that biogeochemical interactions influence the distribution of chemicals in an estuary and that elemental distribution is not solely the product of mixing processes. The sediment/water interface is a particularly important site for reactions to occur and this, as indicated by Dr Dyer, is influenced by estuarine circulation systems and by secondary currents producing ebb and flow channels. In turn the carrying capacity and concentration of suspended matter will vary with depth, over the tidal period and along the estuary, whilst the effects of scour will result in considerable movement of large masses of settled material and its contained organisms.

It is with this physically mobile background that the relatively small numbers of brackish water species of organisms have to contend, and when this is coupled with the instability of flow, level, and salinity caused by the tides and by changes in the downflow of fresh water, such organisms require considerable physiological versatility to survive. As Professor Beadle has pointed out, the evolution of this physiological versatility has enabled marine organisms to invade fresh waters. The type of investigations requiring elucidation in this field will involve the use of modern techniques, and among these is the need to determine the nature of the rapid adaptations that occur in response to salinity changes for both adult and juvenile stages and in the reproductive and non-reproductive phases of individuals of the same species.

The relatively restricted capacity of aquatic organisms to survive in estuaries reduces the numbers of competitors of species which transpire to be economically important. The potential yield of commercially important lamellibranchs may be as high as 1 kg of dry flesh/m^2 over considerable areas of favoured sites, as Dr Walne has shown. The value of estuaries for the aquaculture of shellfish resides in the shelter they provide, especially at the seed stage. Estuaries are also favoured sites in that they reduce the dissipation of planktonic larvae over large areas, so that they tend to concentrate the economically valuable shellfisheries, as at Penclawdd on the Burry Inlet with its high yield of cockles.

Whilst nutrient salts may be high in estuaries they never reach their maximum potential utility, principally because high turbidities restrict high phytoplankton production. Much of the nitrate and phosphate passing out with sewage effluent is wasted. Primitively there is a cycling of plant nutrients from the land to man and back to the land, but in advanced societies the plant nutrients are not returned to the land but to rivers and estuaries and ultimately to the sea, where they become less accessible in the biogeochemical circulation. Seemingly then we are today slowly but surely degrading the fertility of the land surface and although all the fertility is not completely lost, for some is recovered from sewage sludge, the losses are nevertheless substantial. Since we in Britain import half our food supplies this means that we tip away part of the fertility of 25 000 000 acres of Britain as well as 25 000 000 acres of other parts of the world.

Drs Milne and Dunnet dealt with the productivity and the trophic relations of the fauna of an unpolluted estuary and showed that such typically estuarine animals as *Hydrobia, Corophium* and *Mytilus* attained numbers of up to a hundred thousand per square metre. In contrast our observations on *Hydrobia* in the oxygen sag area of the Thames at Greenhithe gave maximum values of 14 000/m^2 at mean tide level and 4 100/m^2 at low water of neaps. But how valid are the comparisons? Since it now appears that British estuaries are going to be the subject of a widescale investigation in the immediate future one of the prerequisites must be the standardisation of statistically significant techniques if we are going to make comparisons between them in terms of cause and effect. It must also be realised that sampling for different animals is likely to involve different procedures, for to use the same methods for sampling burrowing animals (*e.g.* tubificids or nereids) and those that inhabit the surface layers (*e.g.* *Corophium*) would appear to be invalid. The necessity then arises for preliminary fundamental studies of the distribution and seasonal cycles of even our most common species if we are to approach estuarine ecology on a quantitative basis. One of the significant features of Drs Milne and Dunnet's work is that they have undertaken quantification of feeding relationships and overwintering metabolism of the mussels. Moreover, they have related these to the daily and seasonal rhythms and activities of several species of birds and fish.

The only paper on pollution presented at this symposium is that of Mr Barrett, which refers to the dissolved oxygen content of the Thames. Using statistical procedures he discussed the dissolved oxygen content in relation to such factors as fresh water flow, temperature and season, as a result of which long-term changes in the condition of the Thames have been analysed. His results have been arrived at by averaging the dissolved oxygen content for three monthly periods and he states that due to relatively recent improvements in the treatment of London's sewage the whole estuary has now been aerobic for several years and there has been a return of fish life. Nevertheless, there still exists an oxygen sag of no mean significance between London Bridge and Greenhithe, and for the past three years we have sampled at fortnightly intervals along this sag for fish and shrimps, benthos and plankton. The recent sewage strike provided us with a natural experiment, and to capitalise on this we sampled for fish and shrimps at weekly intervals. The oxygen levels reached during this strike were very low and correlated with this was an almost complete absence of fish and shrimps off West Thurrock, where over previous years we had established an overall seasonal pattern of winter catches (Huddart and Arthur, 1971), punctuated by drops in fish and shrimp numbers associated with intermittent periods of oxygen depression.

The paper by Dr Jefferies deals with salt-marshes in relation to inorganic nutritional problems. He considers that salt-marshes may be large sinks which contain an abundance of common inorganic ions that are cycled in the environment as a result of the various geochemical processes taking place. The exceptions to this are that nitrogen and phosphorus, which, though often in short supply in salt-marshes particularly in the upper reaches, are strongly linked to

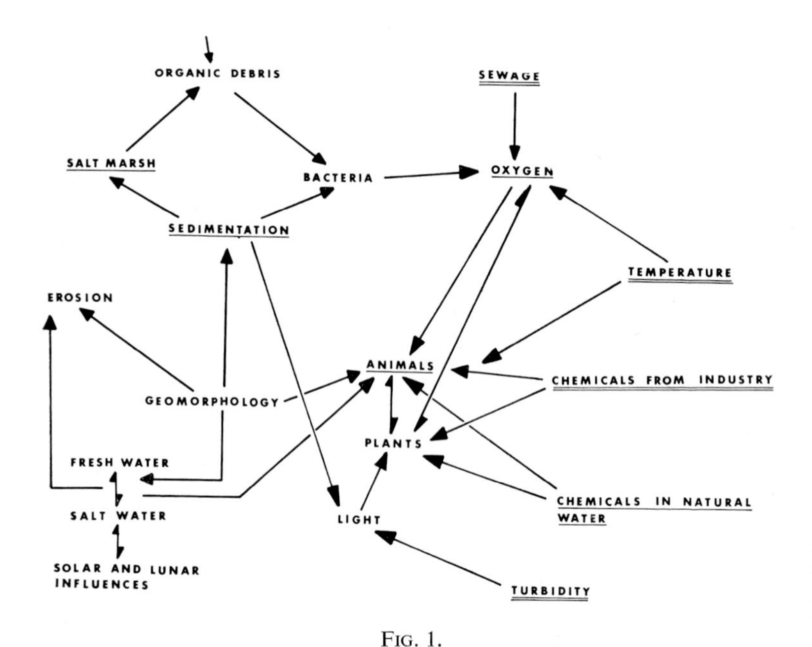

FIG. 1.

the biological cycles occurring within this type of habitat. There is considerable point to point variation in respect of salinity, water level and oxygen which produces a complex mosaic of vegetation where the classical macrozonation of salt-marshes is not clearly discernible. Extreme heterogeneity and genetic variation is common amongst salt-marsh populations but the selective advantage conferred is not altogether clear. Phenotypic plasticity at the morphological and physiological levels is also evident and Dr Jefferies suggests that this is a reflection of the high degree of 'environmental uncertainty'.

Each of the papers shows a highly specialised approach to subjects within the worker's own field of activity, and on a breakdown we have had three papers dealing with estuarine animals, one on sedimentation, one on chemistry, one on pollution and one on plant ecology. If one is bluntly honest as a result of listening to these well prepared and presented papers, those which do not fall within the purview of one's own specialisation are particularly difficult to understand and to interpret in relation to one's own approach to estuarine studies. The approach for future investigations must take this into account. An over-simplified diagram, based on a mechanical analogue described elsewhere (Arthur, 1969), however, indicates the necessity for a multidisciplinary attitude towards estuarine work. The elements underlined with a single straight line in this diagram (Fig. 1) refer to the subjects discussed in this symposium and those with a double straight line to introduced components or to elements subject to change induced by human influence. Thus it is clear that many links in the chain require to be forged, and whilst specialists serve to give unidisciplinary information, cross-fertilisation across boundaries becomes essential if we are to comprehend the estuarine ecosystem as a whole. The need then is for hydrographers, biologists, sedimentologists, chemists, bacteriologists, engineers and representatives of other disciplines to integrate their research and to discuss current problems of both practical and academic importance. For this reason I would strongly support the proposition that deletion of the word 'biological' from the name of the Association is necessary forthwith if it is to have the initial impetus to make the association a 'going concern'.

REFERENCES

Arthur, D. R. (1969). *Survival: Man and his environment.* London: English Universities Press.

Huddart, R. and Arthur, D. R. (1971). 'Whitebait and shrimps in the polluted Thames estuary.' *Int. J. environ. Studies*, **2**, 21–34.

Index